STATE MACHINES
IN
VHDL
Root Functions
Vol. 5

State Machine Design for Arithmetic Processes

Daryl Ray Hawkins

Revision Dates
July 2018 First Release

Table of Contents

Overview

Roots, technically referred to as radicals, are the inverse of raising a number to a power. The most common of these radicals are square and cube root functions and are found in almost all disciplines using mathematical expressions. Higher orders like fourth root, fifth root and so on, have very limited applications. While a simple square root algorithms are very similar to division the higher roots are not, but most maintain a remainder.

Chapter 2 provides the relevant Rules/Laws to refresh the reader, in a distilled form, which govern radicals. General academic information is limited because the purpose of this book is to provide hardware techniques to design radical functions, not provide a comprehensive study of the subject.

Five designs are provided in this book. Square root has two, one is very efficient in terms of resource use and the second is adjoined with specialized circuits to provide high performance. Likewise, there are two cube root functions, one more efficient and one high performance. The fifth and final design supports a scalable radical with an index (order) input parameter ranging from 2nd to the 5th order utilizing the Newton-Raphson method.

All implementations are fixed-point and scalable. Volumes 2 and 3 of this book series detail multiplier and divider modules that can be used as support components for the fifth design, or the reader can implement their own. Performance is dependant on the capability of the external modules selected.

Copies of all source code, in text file form, used in this book is available for purchased through the web site listed below -- under publications, or pubs.

http://www.hawkinseng.com

1   Prerequisites

State Machine techniques used throughout this book are covered in the "STATE MACHINES IN VHDL *Composition* Vol. 1" book, which is a prerequisite, and should be reviewed by the reader.

Additionally, the "STATE MACHINES IN VHDL *Multipliers* Vol. 2" and "STATE MACHINES IN VHDL *Dividers* Vol. 3" contain useful information on formatting and rounding techniques. These subjects will not be repeated here.

Reviewing the IEEE 754-1985/2008 Floating-Point specifications is also recommended.

2 Rules/Laws for Radicals

2.1 Definition

The term radical comes from the word radix, which means root. Root and radical are used synonymously in this book to represent the radical function – also being the inverse function of raising a number to a power. Figure 1 shows the anatomy of a root function as normally depicted.

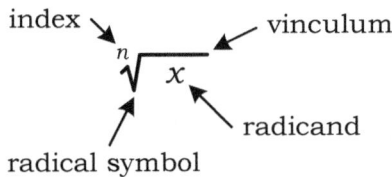

index   vinculum

radicand

radical symbol

Figure 1

- Index

Represents the order of or n^{th} root and is always a positive integer greater than 1. An index of 2 denotes a square root, 3 a cube root, 4 a forth root, 5 a fifth root and so on. The index must be explicitly written except if the function is a square root, in which case the index is absent or implied. When present, it should be positioned over or near the V-portion of the radical symbol.

- Radical Symbol

Means "root of", which is similar to a check mark. Expressions directly under the vinculum (line), the radicand, is operated on. If the vinculum is not present, only the variable just to the right is operated on, unless an expression is within parentheses. See radicand.

- Vinculum (optional)

Is the horizontal line directly over the radicand and attached to the radical symbol. It must be long enough to extend over the entire expression.

- Radicand

The expression operated on or the operand of the radical function.

Caveat: *All designs within this book require that the radicand be reduced to a single fixed-point number or variable.*

2.2 Rules of Combining Radicals

$$\sqrt[n]{x} \bullet \sqrt[n]{y} = \sqrt[n]{(xy)}$$

$$\sqrt[n]{(x \backslash y)} = \sqrt[n]{x} / \sqrt[n]{y}$$

$$(a \bullet \sqrt[n]{x}) + (b \bullet \sqrt[n]{x}) = (a + b) \bullet \sqrt[n]{x}$$

$$\sqrt[m]{(\sqrt[n]{x})} = \sqrt[mn]{x}$$

2.3 Relationship to Exponents

$$\sqrt{x} = x^{\frac{1}{2}}$$

$$\sqrt[n]{x} = x^{1/n}$$

$$\sqrt[n]{(x^m)} = (\sqrt[n]{x})^m = x^{m/n}$$

2.4 Reduction and Simplification

- Radicand must be reduced to a single variable with no exponent/powers.
- Fractional radicands must be represented by a single fixed-point number.
- Index should be as small as possible, no reduction left.

$$\sqrt{(x^2 \bullet y)} = x \bullet \sqrt{y}$$

$$\sqrt[n]{(x^n \bullet y)} = x \bullet \sqrt[n]{y}$$

2.5 Index and Sign of Radicand

Degree of Root (n)	Sign of Radicand	Results
Even index	+	Positive result
Odd index	+	Positive result
Even index	–	There is no possible answer. An example would be taking the square root of a negative number. The result would be positive with an associated imaginary number $i = \sqrt{-1}$. See below.
Odd index	–	Negative result

2.6 Real, Imaginary, and Complex numbers

Real numbers are those containing all rational and irrational numbers. Imaginary numbers are those that are outside of real numbers. The primary example is when taking the square root of a negative number. There is no possible result in real numbers, because the square of a negative number is positive.

$$\sqrt{-x} = \sqrt{-1} \bullet \sqrt{x} = i \bullet y$$

The solution is to remove the negative sign from the radicand and take the root of its positive equivalent, then associate it with $\sqrt{-1}$, which is represented by the imaginary i unit. Complex numbers combine real and imaginary components. Operations are governed by certain rules and are handled at the higher levels of the mathematical expressions in which the radical resides.

> Note: *All designs in this book that support an even index have an output parameter of i to indicate the presence of $\sqrt{-1}$ associated with negative radicands.*

3 Fixed Versus Floating-Point

All operations are inherently fixed-point and are favored over
floating-point, primarily because of the additional overhead burden
associated with floating-point normalization (required implied 1
format). The down side to fixed-point is the reduced accuracy when
approaching the lower boundaries, and the magnitude
confinements of its explicit range.

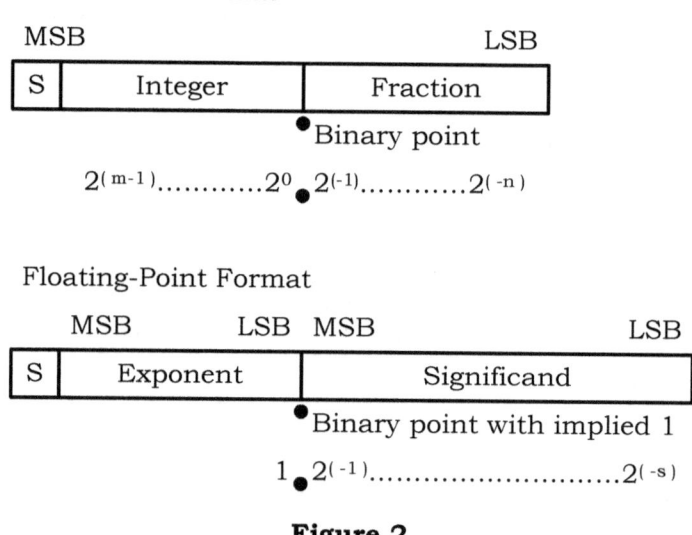

Fixed-Point Format

Floating-Point Format

Figure 2

Fixed-point is a two's complement binary number with a fixed
number of integer bits and a fixed number of fractional bits with
the binary point being implicit between the two. The industry Q
notation for fixed-point binary numbers is Qm.n, 'm' representing
the number of integer bits excluding the sign bit; 'n' representing
the number of fractional bits.

Note: *An example for Qm.n would read Q15.16 for a 32-bits.*

Floating-point is a signed magnitude number with an implied
binary point. The significand (all bits to the right of the binary
point) is left-shifted so that the most significant '1' bit is positioned
to the left of the implied binary point and discarded (not saved),
while offsetting the exponent (bias) accordingly.

Two of the most common implementations are single and double precision, 23-bit and 52-bit significand (mantissa) excluding the implied 1; 24-bit and 53-bit including it.

The number of bits allocated for the exponent is 8-bits for single precision and 11-bits for double. Their nominal value is an offset from 0, which is +127 and +1023, respectively. Because they are offset values instead of absolute values, they are referred to as a bias instead of an exponent; sometimes called an exponent bias.

> Note: *All implementations in this book are done in fixed-point. Where warranted, notes are provided on floating-point or exceptions for consideration by the designer.*

Notes on Multipliers

Associated extended formats are not needed with multiplication. Extended means that more bits are used during the operational stage than are kept after normalization. The initial product is larger than the normalized product. These lower bits are generally used for rounding, in particular the sticky bit.

Notes for Dividers

Associated extended formats are useful for division. Extended means that more bits are used during the operational stage than are kept after normalization. Unlike multiplication there are no lower unused bits. Extending the quotient provides more lower bits for rounding, in particular the sticky bit.

4 Limiting Factors

4.1 Carry Propagation

Chapter 3 of both "STATE MACHINES IN VHDL Multipliers Vol. 2" and "STATE MACHINES IN VHDL Dividers Vol. 3" address limiting factors, in particular – carry propagation – and should be reviewed by the reader. Root extraction is similar to division because of the presence of a remainder.

Signed-digits, carry-save form, carry-select adders, carry-save adders, and multi-level carry-save adder trees are employed in this book where needed to improve performance. Specific subsections are provided for further clarification. Also see chapter 9.

4.2 Integer Size

There is no inherent limit to the size of an adder in VHDL. There are, however, limitations in representing and monitoring large fixed-point numbers in source code.

Generally, designers use conversion functions to convert between real-number data objects and binary numbers. Real signals or variables can be expressed textually in the floating-point style. This is useful when assigning constants in the source code and for monitoring objects during simulation.

The exponent (**) operator and CONV_INTEGER (std_logic_arith.vhd) or TO_INTEGER (numeric_std.vhd) functions are limited to a range between plus and minus 2147483647 (2147483648), reducing the conversion range of the integer portion to 32-bits (sign included) and the fractional portion to 31-bits (no sign), each.

> Note: *The addendum provides functions that will convert from fixed-point to real-numbers (simulation floating-point data objects) that extend past the integer limitation of VHDL. They may also be used to assign constants with floating-point text representation in synthesizable source code.*

5 Normalizing, Rounding, and Bounds

Simple root extraction closely resembles a division operation, primarily because there is a remainder. The resulting number of bits after extraction, and in some cases reduced extraction time, are unlike multiplication and division.

Typical root functions operate on pairs or groups of bits depending on the root order. Radix of the design and the number of bits or signed-digits retired each iteration are also factors. For example, a square root function evaluates a pair of bits and retires a single bit (radix-2), passing the remainder to the next iteration to be combined with the next two bits of the operand. This means the size of the result will be approximately half the number of bits compared to the input operand. This last point is normally true and can be used to reduce extraction time. Designs in this book extend the number of iterations when generating the fractional portion of the result in order to work down a non-zero or non-terminating remainder. The integer portion is also maintained only because pre-shifting would be required anyway to locate the leading pair of 1-bits.

> Note: *how bits are grouped and why is covered within each particular design description. Only generalizations are used in this chapter. A rule of thumb for the most simple of designs is to divide the number of integer and fractional bits by the root order.*

As for overflow, technically it cannot occur and underflow can only occur if the radicand is too small. See sections *5.3 Rounding* and *5.5 Side Effects* for both fixed-point and floating-point designs. As with other designs in this book series the GRS paradigm is used for rounding. Two extra bits, guard (G) and round (R), are generated during extraction and the remainder creates the sticky bit (S).

> Note: Also, refer to "STATE MACHINES IN VHDL *Multipliers* Vol. 2" and "STATE MACHINES IN VHDL *Dividers* Vol. 3" sections 4.3 and 4.4 each for additional information on rounding bits, modes, and the sticky bit. Exceptions regarding roots are provided at the end of this chapter.

Advanced Algorithms

The last and most advanced design(*Nth Root*) in this book does not use the concept of grouping bits except when scaling. Instead it maintains intermediate data and results within given Qm.n formats so that no normalization is required.

5.1   Fixed-Point

Figure 3 shows the steps for normalizing the outcome for root extraction using fixed-point. The initial root contains a resulting number of integer bits , which may be smaller than the input operand's format, a full number of fractional bits (same as input operand's format), plus two bits for rounding. Position of the binary point is offset from the right by the number of fractional bits plus two. The integer portion and sign are just to the left of that.

Note: *the term "resulting" means the resulting number based on the particular design.*

Rounding, if enabled, combines rounding bits Guard (G) and Round (R) with the remainder Sticky (S). If qualified, 1 is added to the resulting root value. If required by the design, the output is sign extended to fit the input operand's format.

Note: *Most algorithms only extract roots of positive numbers. Typically, negative radicands are first converted to positives, then processed. If a negative result is required the rounded root is converted to a negative. Where applicable the imaginary bit is used to represent negative roots, in particular square roots.*

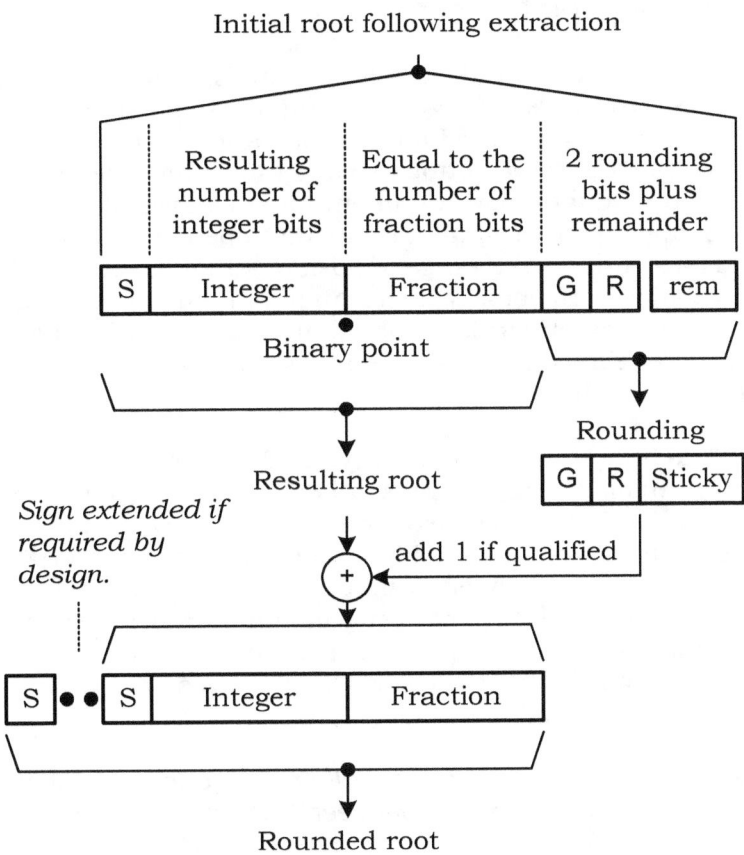

Figure 3 Fixed-Point Normalizing

5.2 Floating-Point

Using floating-point for radicals is different from multiplication or division because there is only one operand instead of two. Also, the implied 1 must be reinserted into the operand and moved so that its position allows the resulting exponent to be divisible by the root's order. A square root's exponent must be divisible by 2, cube root by 3, and so on. Most algorithms that operate on fractional operands account for the range of positions that the leading bit can occupy relative to the binary point. After extraction the exponent is divided by the root order: by 2 for square roots, by 3 for cube and so on. All are relative to the starting bias, 127 for single precision and 1023 for double.

As with fixed-point the number of cycles depends on the root order and bit grouping. Operand size is either 24-bits for single precision and 53-bits for double, plus extended bits, plus two rounding. IEEE stipulates 8 or more extended bits for single precision and 11 or more for double for extended formats, totaling 32+ and 64+, respectively. Some white papers suggest the number of extended bits be equal to the precision itself, resulting in an initial quotient of (2 x 24)+2 and (2 x 53)+2. These number sizes are the same as those used in division.

> Note: *extending formats is not really required for roots because of previously mentioned ratios between radicand and root.*

The extra bits used in extended formats improve the accuracy of the final result after rounding and prevent truncation if the initial root is a very small.

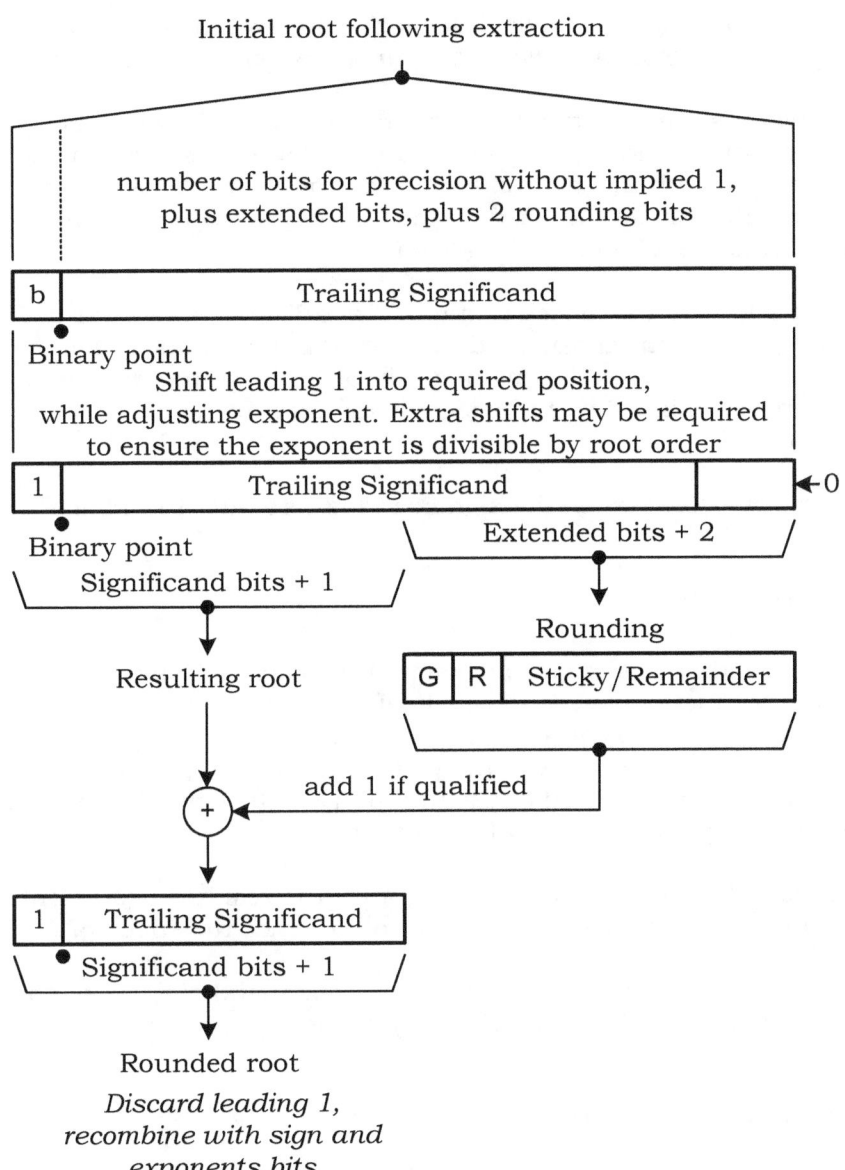

Figure 4 Floating-Point Normalizing

Figure 4 illustrates the steps for finalizing the root from its initial state. The initial root value is shifted, while adjusting the exponent, until the most significant 1 is to the left of the binary point.

Note: *some algorithms operate on the implied 1 being to the right of the binary point within in the range previously mentioned.*

The two bits remaining in the lower half of the trailing significand serve as guard and round. Any remaining lower bits can be used in conjunction with the last remainder to produce the sticky bit. Rounding is further qualified by the LSB of the resulting root, which is to the left of guard and round.

One is added to the normalized root if qualified and afterwards the leading 1 is again repositioned. Afterwards, the rounded root is stripped of its leading '1', making it implied, then the exponent and sign are recombined with the significand into the floating-point format.

Underflow occurs when the significand is zero after rounding.

5.3 Rounding

While rounding is similar to division, it is simpler. The digit recurrence methods used in this book series extract roots as positive numbers. Typically the remainder is positive hence a non-zero remainder results in a sticky bit of 1, otherwise zero. This combined with the guard and round bits and by employing round-toward-nearest-even, generates rounding.

In cases where the remainder is negative the result is decremented by 1 and the remainder is not used to influence the sticky bit.

Advanced algorithms use external multipliers and dividers so that rounding is controlled within those modules.

5.4 The Sticky-Bit

The sticky bit is set in response to a non-zero remainder in fixed-point implementations and can also be used in conjunction with the extended bits in floating point implementations. This last point is not a requirement. The extended bits may be used instead of the remainder.

In algorithms that generate the result and the result minus 1 the sign of the reminder selects between the two possible results, sticky is considered zero when the result minus 1 is selected.

5.5 Side Effects

When taking the root of an integer only value the result is always less in magnitude than the operand, except when taking the root of exactly 1 which is 1. Similarly, when taking the root of a number with both integer and fractional content, the magnitude of the result is still less than the input operand. Roots of fractional only content instead produce a result greater in magnitude, but never reach or exceed 1, even when rounding is occurs.

> Warning: *this last point is not implemented in designs of this book but can be with extra cycles or by using larger operands. Otherwise rounding can result in a 1.0 value.*

Large round-up bias can occur when the Qm.n fractional portion is small. To mitigate this so some designs have a dynamic round control that can be enabled or disabled on a operation by operation basis.

There are cases when rounding a square root, then squaring it, the result will be greater than the original radicand.

6 Square Root Functions

Square roots ($\sqrt{x}$) are the most common of all radical functions. Because of this two designs are provided in this chapter that only extract square roots. One highly efficient design using a shift and subtract algorithm that is very similar to simple sequential division, and a second higher performance design that retires 2-bits per iteration, referred to as radix-4.

While there are some similarities between the two, like employing iterative shift and subtraction cycles or digit-recurrence, bit grouping is different and the radix-4 design uses signed-digit representation and faster adders. Efforts in implementing the latter are well worth it due the reduction in carry propagation, which results in higher achievable clock rates.

Note: *Both designs have an imaginary output bit. If the operand is negative, the imaginary bit will be set true and the radicand is processed as a positive number.*

6.1 Simple Square Root

This design is scalable in terms of the number of integer and fractional bits (Qm.n fixed-point format). The square root algorithm in this section requires that the operand be an even number of bits. Because of this the input radicand may need adjusting by padding the either the integer portion or fractional portion, or both.

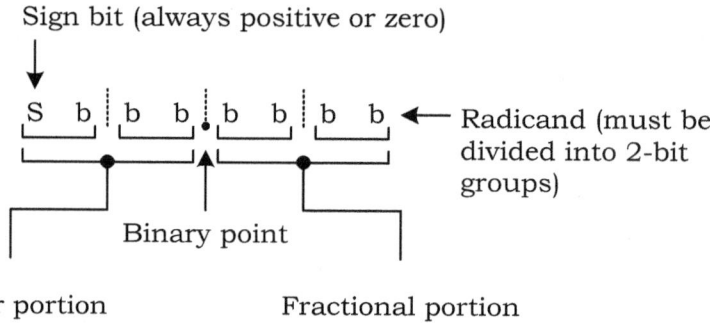

Sign bit (always positive or zero)

S b b b b b b b ← Radicand (must be divided into 2-bit groups)

Binary point

Integer portion   Fractional portion

Figure 5

Working Radicand register:

Figure 5 illustrates an ideal operand with an even number of bits, partitioned into 2-bit groups. When the selected Qm.n format for the input radicand is not ideal, the following rules apply during instantiation when creating the size of the working radicand register:

- Integer portion – if the number of integer bits is odd, use the operand's integer size for the working radicand because the sign bit completes the final pair. If even, extend the MSB by 1-bit resulting in a zero position to the left of the sign bit to complete the upper pair.
- Fractional portion – if the number of fractional bits is even, use the operand's fractional size for the working radicand because the size is already even. If odd, add a bit position to the right of the LSB.

Note: *Adding these bits is referred to as padding.*

During each cycle the working radicand is left-shifted by two bits with zeros replacing the lower two positions. See Figure 7.

Working Root register:

The working root register is sized differently and in some cases requires the MSB be padded by an additional bit to insure the final MSB is zero, even though the size of the working radicand was already padded. Rules for sizing the working root are as follows:

- Integer portion – if number of bits in integer portion of the input operand is odd, integer portion is padded by 1-bit position to the left of the MSB.
- Fractional portion - same as the input operand's plus two bits for rounding, guard and round.

Note: *The sign bit need not be added because it is factored into the integer portion of the working root register. A leading zero in the result is always guaranteed.*

Steps in Algorithm:

Figure 6 depicts the algorithm from a conceptual point of view.

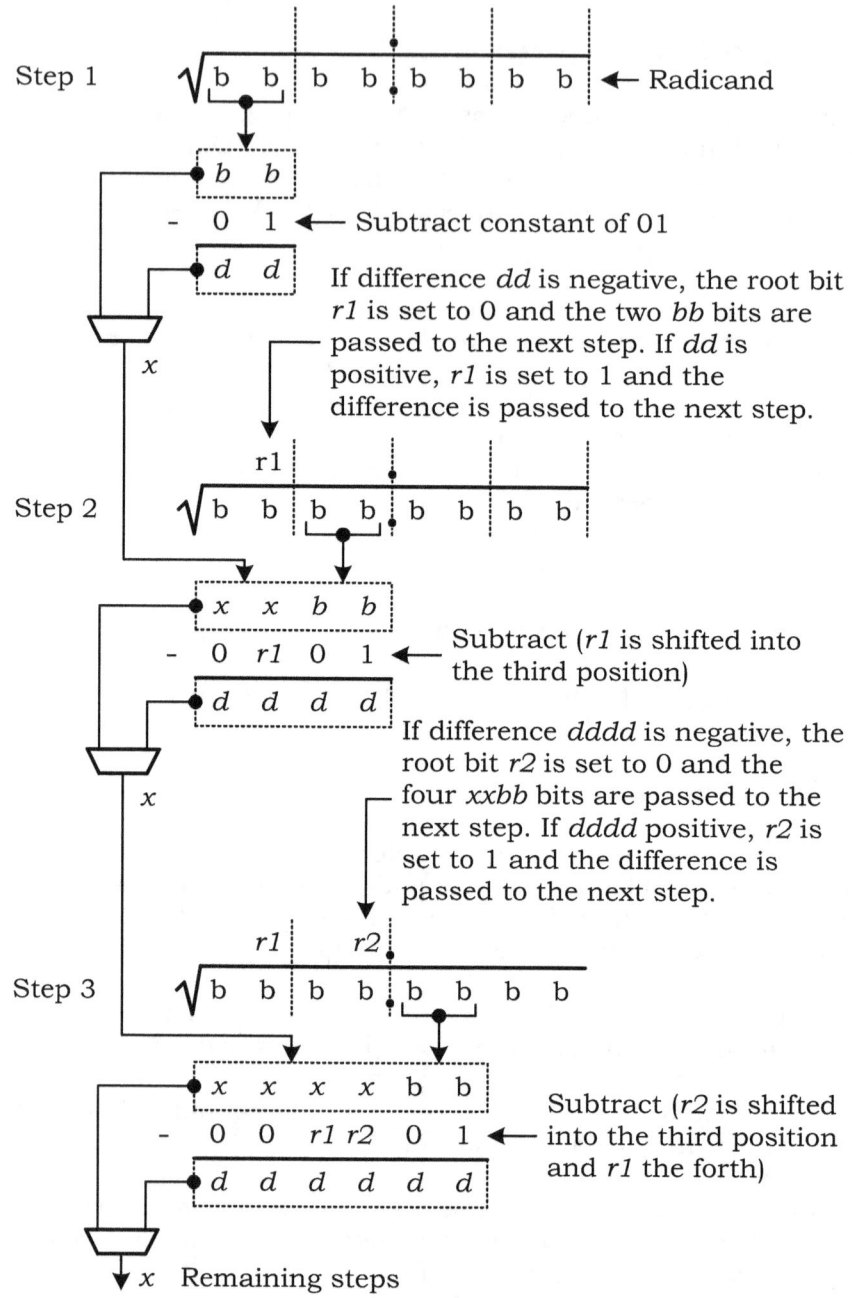

Figure 6

Design Flow:

Circuit design of the algorithm is shown below in Figure 7. First the working radicand is sized and initialized as stated earlier in this section, with the remainder and working root cleared.

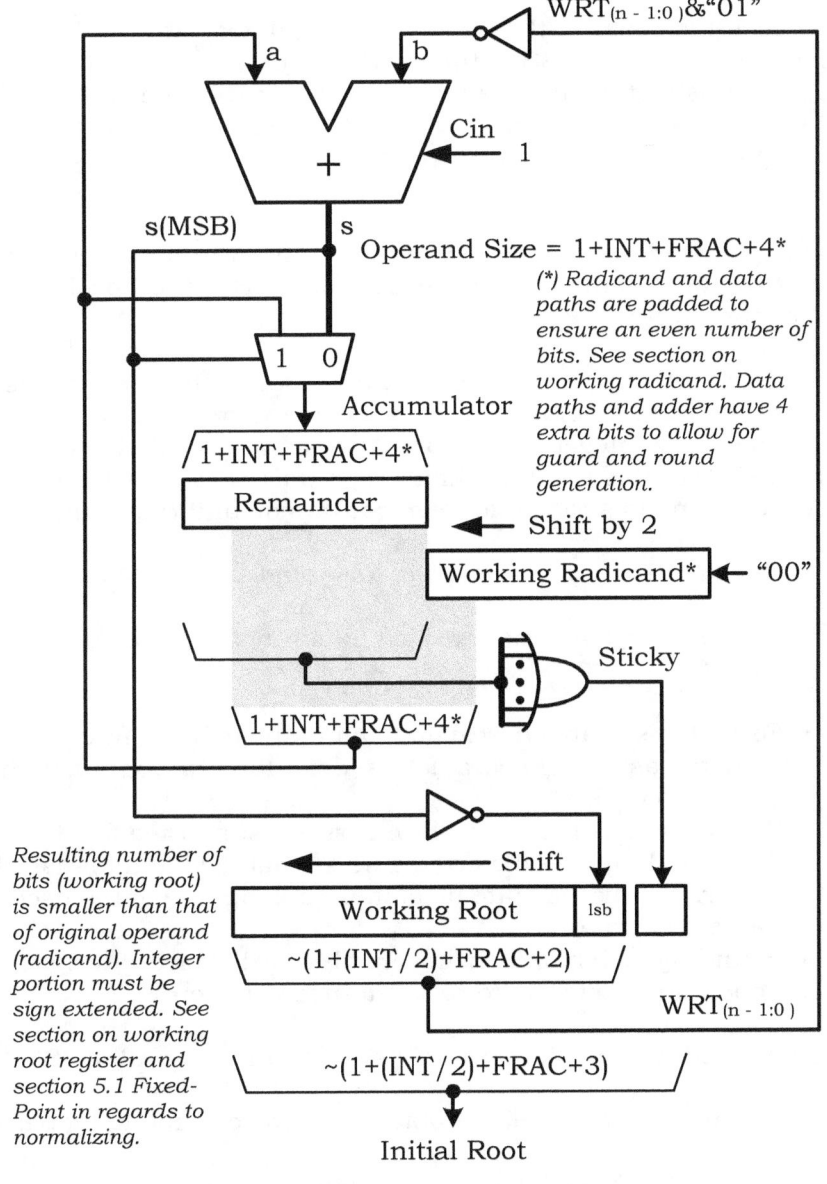

WRT$_{(n - 1:0)}$&"01"

a   b

Cin   1

+

s(MSB)   s

Operand Size = 1+INT+FRAC+4*

(*) Radicand and data paths are padded to ensure an even number of bits. See section on working radicand. Data paths and adder have 4 extra bits to allow for guard and round generation.

1   0

Accumulator

/ 1+INT+FRAC+4* \

Remainder

Shift by 2

Working Radicand* ← "00"

Sticky

\ 1+INT+FRAC+4* /

Resulting number of bits (working root) is smaller than that of original operand (radicand). Integer portion must be sign extended. See section on working root register and section 5.1 Fixed-Point in regards to normalizing.

Shift

Working Root   lsb

\ ~(1+(INT/2)+FRAC+2) /

WRT$_{(n - 1:0)}$

\ ~(1+(INT/2)+FRAC+3) /

Initial Root

Figure 7

The upper 2-bits of the working radicand are combined with the remainder (minus its upper 2-bits) creating the '*a*' input into the adder. The working root appended by bits "01" and inverted creates the '*b*' input to the adder. Resulting in a two's complement subtraction.

If the result is positive, the remainder is updated with the difference '*s*' and the LSB of the working root is set to '1'. If negative, it is instead updated with the '*a*' input to the adder and the LSB of the working root is set to '0'. During each cycle the working radicand is left shifted by two bits and the working root by one.

Finally, the working root is mapped back into the output's Qm.n format zeroing the upper bits between the MSB of the working root to the root output.

The number of clock cycles required to complete the operation is equal to the number of bits in the working root register. This is the input operand's integer size/2, plus the operand's number of fractional bits, plus two rounding bits, plus 4 clocks for pre and post processing to account for sign conversion and rounding.

Note: *See chapter 5 for Normalizing, Rounding, and Bounds.*

The example state machine provided illustrates the *Simple Square Root.*

- Format is scalable fixed-point Qm.n, integer and fractional bit lengths as generic parameters, defaulting to Q31.32, with a range to Q63.64.
- Signed two's complement numbers are supported for the radicand input. All conversions are handled and managed internally by the state machine. An imaginary bit is used for a negative root.
- Rounding is dynamic, being enabled or disabled during each operation. Defaults to round-to-nearest-even.

As configured for Q31.32, test builds were run using Xilinx ISE.

The following performance was obtained with corresponding parts.

Xilinx Spartan XC3s500e greater than 114MHz
Xilinx Virtex XC5vlx30 greater than 232MHz

```
-------------------------------------------------------------------
--   SimpleSqrt.vhd (Simple Square root)
-------------------------------------------------------------------
library IEEE;
use IEEE.std_logic_1164.all;
use IEEE.numeric_std.all;
-- user packages

entity SimpleSqrt is
--
--   Qmn fixed point format is used.
--
--        sign         binary point
--          |            |
--   format <s>(integer bits).(fractional bits)
--          _____/ _____/
--          INT_SIZ        FRAC_SIZ
--
--
generic (INT_SIZ: integer range 0 to 63 := 31;
         FRAC_SIZ: integer range 0 to 64 := 32);
port
(
    clk,rst: in std_logic; -- system clock and reset (synchronous)
    -- inputs
    rnd_en: in std_logic; -- enable rounding for current cycle
    extr: in std_logic; -- initiate root extraction
    rcd: in signed((1+INT_SIZ+FRAC_SIZ)-1 downto 0); -- radicand
    -- outputs
    rdy: out std_logic; -- data ready
    udr: out std_logic; -- underflow
    i: out std_logic; -- imaginary
    rt: out signed((1+INT_SIZ+FRAC_SIZ)-1 downto 0) -- root
);
end SimpleSqrt;

architecture RTL of SimpleSqrt is
----------------------
-- General Functions
----------------------
-- pad even size by 1
function PadEvenByOneBit(siz: integer) return integer is
begin
    if(siz mod 2) = 0 then -- even, pad by 1-bit
```

```
        return(siz+1);
    else -- odd, leave as is
        return(siz);
    end if;
end function;
-- pad odd size by 1
function PadOddByOneBit(siz: integer) return integer is
begin
    if(siz mod 2) /= 0 then -- odd, pad by 1-bit
        return(siz+1);
    else -- even, leave as is
        return(siz);
    end if;
end function;
-- compute working radicand integer size (see description below)
function ComputeWrtIntegerSiz(siz: integer) return integer is
begin
    if (siz mod 2) = 0 then
        return((siz+2)/2); -- (int size) + 1 + S
    else
        return((siz+3)/2); -- (int size + 1) + 1 + S
    end if;
end function;


-----------------------
-- declared constants
-----------------------
-- working radicand size, pad to 2-bit boundaries (sign factored in)
constant WRCD_INT_SIZ: integer := PadEvenByOneBit(INT_SIZ);
constant WRCD_FRAC_SIZ: integer := PadOddByOneBit(FRAC_SIZ);
constant WRCD_SIZ: integer := 1+WRCD_INT_SIZ+WRCD_FRAC_SIZ;

-- internal data path size (extra 4-bits for guard and round generation)
constant D_SIZ: integer := WRCD_SIZ+4;
```

-- The ratio of significant bits between the working radicand
-- and the working root register is 2:1. When creating the
-- working radicand constants (above), integer and fractional
-- lengths were adjusted to guaranteed an even number of bits
-- or pairs for the extraction algorithm.
--
-- The working root register is scaled somewhat differently. If
-- INT_SIZ is odd, an extra bit position is added to ensure the
-- the final MSB position is zero. As for the fractional portion,

```vhdl
-- its length is the same as that of the FRAC_SIZ to have the
-- greatest accuracy while reducing round-up errors.
--


-- working root integer size (extra MSB included)
constant WRT_INT_SIZ: integer := ComputeWrtIntegerSiz(INT_SIZ);
-- fractional portion
constant WRT_FRAC_SIZ: integer := FRAC_SIZ;
--include guard and round bits
constant WRT_SIZ: integer := WRT_INT_SIZ+WRT_FRAC_SIZ+2;

-- constants for normalized root (defaults to output port pins)
constant NRT_INT_SIZ: integer := INT_SIZ;
constant NRT_FRAC_SIZ: integer := FRAC_SIZ;
constant NRT_SIZ: integer := 1 + INT_SIZ + FRAC_SIZ;


---------------------
-- declared signals
---------------------
signal wrcd: signed(WRCD_SIZ-1 downto 0) := (others=>'0');
signal a: signed(D_SIZ-1 downto 0) := (others=>'0');
signal b: signed(D_SIZ-1 downto 0) := (others=>'0');
signal sum: signed(D_SIZ-1 downto 0) := (others=>'0');
signal remainder: signed(D_SIZ-1 downto 0) := (others=>'0');
signal wrt: signed(WRT_SIZ-1 downto 0) := (others=>'0');
signal grs: unsigned(2 downto 0) := (others=>'0');
signal cnt: integer range 0 to (wrt'length-1) := 0;
signal busy: std_logic := '0';
signal rnd: std_logic := '0';


----------------------
--   enumeration lists
----------------------
type sm_def is
(
    RESET,
    START_EXTR,
    PROCESS_SIGN,
    EXTRACT_ROOT,
    ROUND,
    ROUND2,
    ROUND3
);
signal state: sm_def := RESET;
```

------------------------------- module code ----------------------------

```vhdl
begin
-------------------------------------
--   Simple Square Root state machine
-------------------------------------
process(rst,clk)
begin
    if(rst='1') then
        -- outputs
        busy<='0'; udr<='0'; i<='0'; rnd <= '0';
        rt <= (others=>'0');
        -- local
        remainder <= (others=>'0');
        wrt <= (others=>'0');
        grs <= (others=>'0');
        -- states
        state <= RESET;
    elsif rising_edge(clk) then
        --
        --   state machine body
        --
        case state is
            -- reset state
            when RESET =>
                state <= START_EXTR;
            -- wait for signal to start extraction
            when START_EXTR =>
                if(extr = '1') then
                    busy<='1'; udr<='0'; i<='0'; rnd <= rnd_en;
                    -- account for padding in integer portion by sign extending
                    wrcd(wrcd'high downto (wrcd'high - WRCD_INT_SIZ)) <=
                resize(rcd(rcd'high downto (rcd'high - INT_SIZ)),WRCD_INT_SIZ+1);
                    -- account for padding in fractional by appending a 0   to content
                    if(FRAC_SIZ > 0) then
                        -- append a zero if FRAC_SIZ is odd
                        if ((FRAC_SIZ mod 2) /= 0) then
                            wrcd(0) <= '0';
                            for i in 1 to WRCD_FRAC_SIZ-1 loop
                                wrcd(i) <= rcd(i-1);
                            end loop;
                        -- copy as is if FRAC_SIZ is even
                        else
```

```vhdl
                for i in 0 to WRCD_FRAC_SIZ-1 loop
                    wrcd(i) <= rcd(i);
                end loop;
            end if;
        end if;
        state <= PROCESS_SIGN;
    end if;
-- convert working register if negative
when PROCESS_SIGN =>
    if(wrcd(wrcd'high) = '1') then
        i <= '1'; -- set imaginary
        wrcd <= (not wrcd) + 1;
    end if;
    -- clear objects
    cnt <= 0;
    wrt <= (others=>'0');
    remainder <= (others=>'0');
    state <= EXTRACT_ROOT;

---------------------------
-- extract root algorithm
---------------------------

when EXTRACT_ROOT =>
    -- left-shift working radicand for next cycle
    wrcd <= wrcd(wrcd'high-2 downto 0)&"00";
    -- check upper bit for negative result
    if(sum(sum'high) = '1') then
        remainder <= a; -- restore remainder left-shifted
    else
        remainder <= sum; -- save difference
    end if;
    -- retire next root bit and left shift
    wrt <= wrt(wrt'high-1 downto 0)&(not sum(sum'high));
    -- sequence counter based on wrt size
    if(INT_SIZ mod 2) = 0 then
        -- even has no extra leading zero bit
        if(cnt < (wrt'length-1)) then
            cnt <= cnt + 1;
        else
            state <= ROUND;
        end if;
    else
        -- odd has extra leading zero bit
        if(cnt < (wrt'length-2)) then
            cnt <= cnt + 1;
```

```vhdl
                else
                    state <= ROUND;
                end if;
            end if;
        ------------------------------------
        -- round and normalize working root
        ------------------------------------
        when ROUND =>
            -- guard and round bits
            grs(2 downto 1) <= unsigned(wrt(1 downto 0));
            if(remainder /= 0) then -- set sticky
                grs(0) <= '1';
            else
                grs(0) <= '0';
            end if;
            state <= ROUND2;
        when ROUND2 =>
            -- round result to nearest even number
            if (rnd = '1' and (grs > 4 or (grs = 4 and wrt(2) = '1'))) then
                wrt(wrt'high downto 2) <= wrt(wrt'high downto 2) + 1;
            end if;
            state <= ROUND3;
        when ROUND3 =>
            -- check for underflow
            if(wrt(wrt'high downto 2) = 0) then
                udr <= '1';
            end if;
            -- properly normalize working root into
            -- output root register. Clear all bit positions
            -- not being assigned during transfer.
            rt <= (others=>'0');
            -- assign significant bits from working root to
            -- output, while removing guard and round.
            for i in wrt'high downto 2 loop
                rt(i-2) <= wrt(i);
            end loop;

            busy <= '0';
            state <= START_EXTR;
        when others =>
            state <= RESET;
    end case;
end if;
end process;
```

```
rdy <= '1' when busy = '0' and (extr = '0') else '0';


------------------
-- adder circuit
------------------
-- remainder+next 2-bits of working register
a <= remainder(remainder'high-2 downto 0)&wrcd(wrcd'high downto wrcd'high-1);
b <= resize(((wrt)&"01"),b'length);
-- subtract b from a via two's complement
sum <= a + (not b) + 1;

end RTL;
```

6.2 Radix-4 Square Root

There are a limited number of papers and textbook examples that provide information on implementing radix-4 square root functions. Publications by Ercegovac and Lang [1] in and around the 1990s and up to a more recent one in 2004 in a book called *Digital Arithmetic* -- is the algorithm of choice and is leveraged here. These writings are difficult to understand and some rely on previous publications when describing internal building blocks and functions. Rather than repeat everything here only specific expressions related to the actual algorithm are referenced, not the supporting arguments or theory. An attempt to use formerly defined expressions, terms, and variable names is made in the event that the reader wishes to further investigate the premises underlying these expressions.

The algorithm is loosely called *continued-sum-recurrence* or *digit-recurrence* and is implemented as radix-4 (two bits are retired each iteration). Conceptually, the upper bits of the residual (remainder) are combined with several significant bits of the result to create an address into a look-up table. Signed-digits {-2,-1,0,+1,+2} comprise the content of the table. Mapping of these digits is the real academic work done by Ercegovac and Lang. The use of carry-save and high-performance adders reduce the effects of carry-propagation thus increases clock speed. Additionally, appendage techniques allow the result (the intermediate root) to grow in length without the use of adders.

Due to the levels of logic required between and within each sub module, an iteration takes two clock cycles instead of one. Since two bits are retired each iteration this design is on par with the *Simple Square Root* in the previous section, in terms of clock cycles, but in terms of clock speed its performance is greatly improved.

While the original algorithm was designed to support fractional operands, normalization and de-normalization of input and output, respectively, were added to support fixed-point operands. This adds a few more states to the design but the performance gains make it viable.

An overview of the architecture is given first followed by expanded explanations of each building block and function.

6.2.1 Architecture Overview

Figures 8 and 9 represent the full design. The radicand input and root output have the same fixed-point two's complement format. If the integer or fractional portions (generics of select Qm.n format) have an odd number of bits, the internal working registers require zero padding to make them even because the algorithm operates on two bits at a time. Mapping between the internal operands and both the radicand and root is done in the first and last steps, respectively. This also properly positions the binary point.

Since the algorithm was designed to operate on fractional numbers only, the leading bits must be positioned to the right of the sign bit which is the MSB of the working register. The requirement of the algorithm is that the operand must be less than 1.0 and greater or equal to 0.01 (¼). The leading bits are located by left shifting the operand by two bits at a time. If one or both bits are 1 the operand is normalized. This value becomes sign extended as a negative number which is the first entry to the ROM table. At this time register F is zero.

Note: *during shifting ncnt (normalizing count) and fcnt (fractional count) are recorded and used later to de-normalize the result.*

The output of the ROM table, $Sj+1$, is in signed-digit form {-2,-,0,+1,+2} and is used to retire the next two result bits as an appendage to registers A and B. Register A is the result (root) in two's complement form and B is the result − 1. Register F is updated at the same time with the previous values of either A or B and an added appendage. During the next clock cycle register F is added to the previous residual x4 (which is in carry-save form) to create the next residual, which is again used to create the next entry into the ROM table, thus the next signed-digit. This continues until the operand has been fully processed.

Having the residual, *ws* (sum) and *wc* (carry), in carry-save form greatly improves performance because carry-save adders have no carry chain. However, the upper eight bit of the *x4ws* and *x4wc* must be added together with a traditional CPA (carry-propagation-adder) to form the upper 7-bits of the ROM table's address. The lower three bits of the address are derived from the upper bits of register A.

In order to improve performance at the same time the address for the ROM table is generated the anticipated values for registers A, B, and F are created based on what the next output of the ROM table might be. Five results are created because the output S_{j+1} could be either -2, -1, 0, +1, or +2. The next S_{j+1} simply steers the corresponding muxes to update registers A, B, and F.

Steps are shown below.

1). Initialization

   $j = 0$ (j is not used in the design but is here for illustration)
   $A [0] = 1.000$---0
   $B [0] = 0.000$---0
   $F [0] = 000.000$---0
   $x4w$ $Reg [0] = 1111.xxx$---x (normalized radicand $-$ 1.0)

2). Generate first S_{j+1}

   $A[j]$ -> ^S |-> ROM table address -> S_{j+1}
   $x4w[j]$ -> ^y |

3). Update registers A, B, and F based on the value of S_{j+1}

   $A[j+1]$ = anticipated value
   $B[j+1]$ = anticipated value
   $F[j]$ = anticipated value
   $x4w$ $Reg[j]$ = $x4w[j]$ (both ws and wc)

4). Combine (add) $x4w$ $Reg[j]$ and $F[j]$ for new residual

   $A[j+1]$ =   -> ^S |-> ROM table address -> S_{j+1}
   $x4w[j+1]$ -> ^y |
   $j + 1$ = j |

5). Update registers A, B, and F based on the value of S_{j+1}

   $A[j+1]$ = anticipated value
   $B[j+1]$ = anticipated value
   $F[j]$ = anticipated value
   $x4w$ $Reg[j]$ = $x4w[j]$ (both ws and wc)

   exit loop or return to step 4

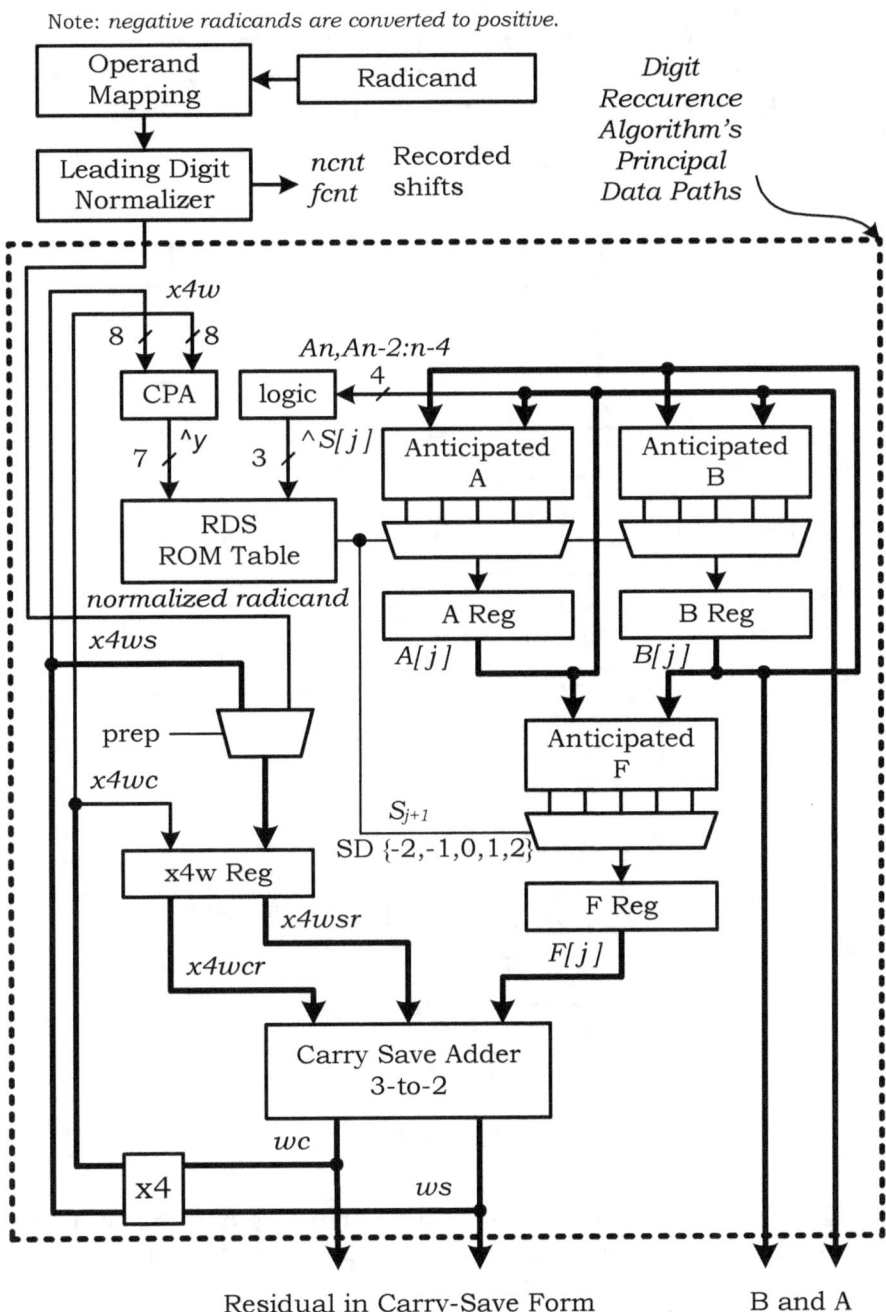

Figure 8

Post processing first involves determining the sign of the last residual which requires *ws* and *wc* to be converted from carry-save form to two's complement through a multi-cycle carry-select fast adder. If the sign is positive register *A* represents the result; if negative, register *B*.

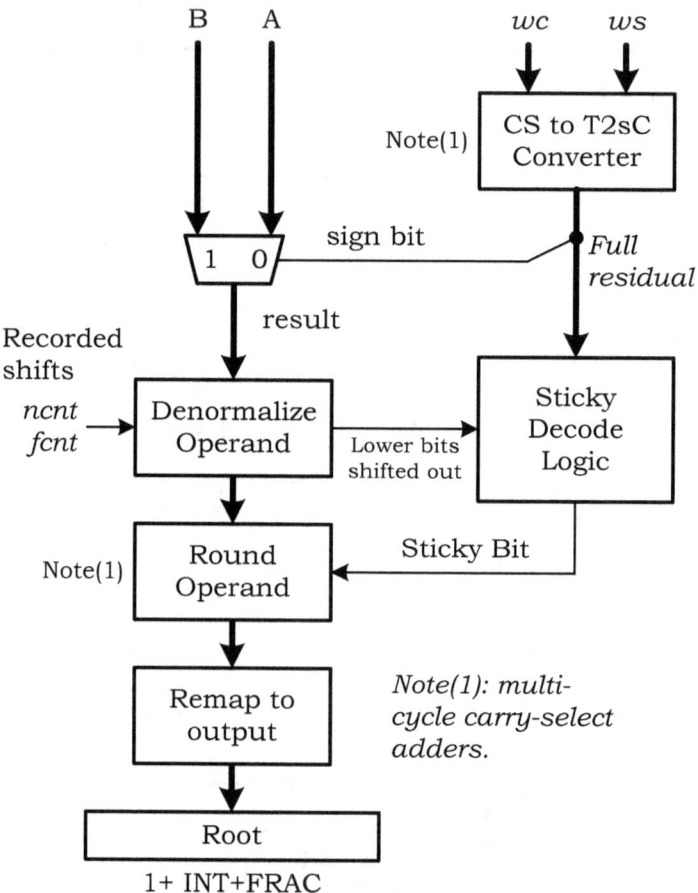

Figure 9

Using the recorded values of *ncnt* and *fcnt* during normalization in conjunction with the input/output fixed-point format constants (generics), the operand is right shifted to properly position the result to where the binary point is for fixed-point. Rounding is qualified by the value in the full residual, any ones shifted out when de-normalizing, and the guard and round bits of the extended internal operand. A multi-cycle carry-select fast adder is used here as well.

The final step involves remapping the internal working register to the output format.

6.2.2 Terms and Variable Names

In the original papers a few symbols are used throughout. Iteration and interval, which are not the same thing, are denoted by the letters j and i, respectively.

Iteration means a specific cycle number. Two 2-bits of the result are retired each cycle. The number of iterations required is approximately the operand size divided by two. In expressions a variable like $A[j]$, the "$[j]$" represents the value of A at the iteration of j. The expression $j+1$ means the value that is posted to the next iteration.

Interval typically means a value or power over a range, i denoting the progression or intermediate steps over that range.

Finally, $S[j]$ is synonymous with $A[j]$, and is retained in a few places to be consistent with previous publications.

6.2.3 Operand Mapping

Operands used in square roots need to be even. When the algorithm is radix-4 the integer and fractional portions each need to be even apart from the sign bit. The design is scalable for both integer and fractional portions through generics but internal working registers must be scaled to account for this even number requirement and the bits from the radicand must be properly placed to maintain the original binary-point.

Working registers account for temporary registers. Principal registers have the same bit padding but may also have additional integer bits between the sign bit and the implied binary point.

Integer portion

If the integer size is an odd number a bit location is inserted between the MSB and the sign bit. If the integer size is zero or even the same value is used for the internal registers.

Fractional portion

If the fractional size is odd a bit location is appended to the LSB. Two extra bits are always added to serve as guard and round bits. The newly extended LSB will be included in the sticky estimate.

6.2.4 Leading Digit Normalizer

The MSB of the internal working registers is the sign bit with the implied binary point to the right of the sign bit. The mapped operand is shifted to the left two bits at a time until one of the two leading bits is true. This exercise normalizes the operand into a fractional number to satisfy the expression below.

$$\tfrac{1}{4} \le x < 1$$

All three of these values qualify 0.01--, 0.10--, or 0.11--. During shifting two counters are advanced. One counts every double shift and the other counts double shifts that are within the fractional portion of the operand. These count values are used to de-normalized the fractional result into the original fixed-point format.

6.2.5 Principal Registers

Registers A, B, F, and $x4wsr/x4wcr$ account for the primary data paths in the algorithm. How and what they are updated with is covered here.

First, expressions for A, B, and F all contain $r^{-(j+1)}$ which in radix-4 is $4^{-(j+1)}$. This represents the lower fractional offset for the appendage of the current iteration. The right most x position of the xx pair.

j = 0	j = 1	j = 2	j = 3
0.25	0.0625	0.015625	0.00390625
S.xx	S.nnxx	S.nnnnxx	S.nnnnnnxx

Registers A and B (internal operand size)

Registers A and B contain the final fractional result in two's complement. On-the-fly-conversion method is employed here. Register B is register A minus the least significant bit of partial result. The $[j]$ represents the value of that register for an iteration of j.

Expressions

$$A[j+1] \quad = \quad A[j] + Sj{+}1 \bullet 4^{-(j+1)} \qquad\qquad \text{if} \quad Sj{+}1 \geq 0$$

$$B[j] + (4 - |Sj{+}1|) \bullet 4^{-(j+1)} \qquad \text{else}$$

$$B[j+1] \quad = \quad A[j] + (Sj{+}1 - 1) \bullet 4^{-(j+1)} \qquad \text{if} \quad Sj{+}1 > 0$$

$$B[j] + (4 - 1 - |Sj{+}1|) \bullet 4^{-(j+1)} \quad \text{else}$$

Technique using appendages, j 0-2 for A and B.

$Sj{+}1$	$j = 0$	$j = 1$	$j = 2$
+2	$A[1] = A[0] + 0.\mathbf{10}$	$A[2] = A[1] + 0.00\mathbf{10}$	$A[3] = A[2] + 0.0000\mathbf{10}$
+1	$A[1] = A[0] + 0.\mathbf{01}$	$A[2] = A[1] + 0.00\mathbf{01}$	$A[3] = A[2] + 0.0000\mathbf{01}$
0	$A[1] = A[0] + 0.\mathbf{00}$	$A[2] = A[1] + 0.00\mathbf{00}$	$A[3] = A[2] + 0.0000\mathbf{00}$
-1	$A[1] = B[0] + 0.\mathbf{11}$	$A[2] = B[1] + 0.00\mathbf{11}$	$A[3] = B[2] + 0.0000\mathbf{11}$
-2	$A[1] = B[0] + 0.\mathbf{10}$	$A[2] = B[1] + 0.00\mathbf{10}$	$A[3] = B[2] + 0.0000\mathbf{10}$

$Sj{+}1$	$j = 0$	$j = 1$	$j = 2$
+2	$B[1] = A[0] + 0.\mathbf{01}$	$B[2] = A[1] + 0.00\mathbf{01}$	$B[3] = A[2] + 0.0000\mathbf{01}$
+1	$B[1] = A[0] + 0.\mathbf{00}$	$B[2] = A[1] + 0.00\mathbf{00}$	$B[3] = A[2] + 0.0000\mathbf{00}$
0	$B[1] = B[0] + 0.\mathbf{11}$	$B[2] = B[1] + 0.00\mathbf{11}$	$B[3] = B[2] + 0.0000\mathbf{11}$
-1	$B[1] = B[0] + 0.\mathbf{10}$	$B[2] = B[1] + 0.00\mathbf{10}$	$B[3] = B[2] + 0.0000\mathbf{10}$
-2	$B[1] = B[0] + 0.\mathbf{01}$	$B[2] = B[1] + 0.00\mathbf{01}$	$B[3] = B[2] + 0.0000\mathbf{01}$

Register F (internal operand size + 2 integer bits)

Expressions

$$F[j] \;=\; -(2A[j] + Sj{+}1 \cdot 4^{-(j+1)}) \cdot \; Sj{+}1 \qquad \text{if} \;\; Sj{+}1 > 0$$

$$(2B[j] + (8 - |Sj{+}1|) \cdot 4^{-(j+1)}) \cdot |Sj{+}1| \qquad \text{if} \;\; Sj{+}1 < 0$$

Technique using appendages, j 0-2 for F

$Sj{+}1$	$j = 0$	$j = 1$	$j = 2$
+2	$F[0] = \bar{a}11.00$	$F[1] = \bar{a}\bar{a}\bar{a}.1100$	$F[2] = \bar{a}\bar{a}\bar{a}.\bar{a}\bar{a}1100$
+1	$F[0] = 1\bar{a}1.11$	$F[1] = 1\bar{a}\bar{a}.\bar{a}111$	$F[2] = 1\bar{a}\bar{a}.\bar{a}\bar{a}\bar{a}111$
0	$F[0] = 000.00$	$F[1] = 000.0000$	$F[2] = 000.000000$
-1	$F[0] = 0b1.11$	$F[1] = 0bb.b111$	$F[2] = 0bb.bbb111$
-2	$F[0] = b11.00$	$F[1] = bbb.1100$	$F[2] = bbb.bb1100$

Note: *$\bar{a}$ is an inverted bit from register A, b is a non-inverted bit from register B. Each bit position represents all active bit positions for that iteration.*

Registers x4wsr/x4wcr (internal operand size + 3 integer bits)

Register *x4wsr* contains the *sum* of the last residual times 4 and *x4wcr* contains the carry component (carry-save form). The upper 8-bits of each are combined with a carry propagation adder to create the upper 7 address bits into the ROM table.

Also, the entire register values are added to the content of the *F* register to generate the next residual, *ws* and *wc*, thus the next ROM table address.

6.2.6 Carry-Save Form

Carry-save form reduces the propagation delay of adding three numbers to a single 1-bit full-adder. Inputs include the last sum and carry of the last residual times 4 (shifted left two places), and the content of the *F* register which is in two's complement. See Figure 8. Also see chapter 9 for more on carry-save adders.

```
x4wsr[j]    Snnn.nnn --- nnn
x4wcr[j]    Snnn.nnn --- nnn
F[j]         Snn.nnn --- nnn
-------------------------------------
             nn.nnn --- nnn    ws[j+1] residual sum
            nnn.nnn --- nn     wc[j+1] residual carry
```

Note: *the upper two bits out of the carry-save adder are discarded. ws and wc* (internal operand size + 1 integer bits).

$$w[j+1] = x4wr[j] + F[j]$$

6.2.7 Signed Digit Representation

Typically signed-digits are encoded with some limited carry information so when combined propagation delays are significantly reduced. Signed-digits are used here to encode the next sub-operation within the algorithm, namely the update of registers A, B, and F based on the current residual and partial result.

The digit set {-2, -1, 0, +1, +2} is each represented by 3-bits. Used in conjunction with on-the-fly conversion carry propagation is eliminated when updating registers A and B. Instead, muxes and appendages are employed.

Also see section 7.2.5 of "STATE MACHINES IN VHDL Dividers Vol. 3".

6.2.8 RDS ROM Table

The acronym RDS stands for Result-Digit-Selection. The resulting signed-digit of the current iteration is the product of the partial residual $^\wedge y$ and partial result $^\wedge S[j]$. The ROM table entries are mapped accordingly.

Boundaries, continuity, and containment are all accounted for in the referenced papers and test books and will not be reproduced here, but the basic expressions and intent are.

ROM Table Address

Signal $^\wedge y$ is produced by adding the 4 integer and four fractional bits of $x4ws$ and $x4wc$ to produce upper 7-bits of address into the ROM table. It is used to address the y-axis in Figure 10.

Note: *the ROM table is arranged to account for two's complement of +/- 8.xxxx of address range.*

$\wedge S[j]$ is only the lower 3-bits of address to the ROM table. Its is used to address the x-axis of Figure 10. The decode logic is described below.

$\backslash \wedge S \backslash (2) <= A0$ or $A2$;
$\backslash \wedge S \backslash (3) <= '1'$ when $A3 = '1'$ or $(A0 = '1'$ and $j > 0)$ else $'0'$;
$\backslash \wedge S \backslash (4) <= A0$ or $A4$;

Note: *here the lower bit position 0 is the highest significant bit.*

Boundaries

The expressions below represent the upper and lower boundaries for each of the signed-digit values across 8 intervals (x axis), Figure 10.The table below shows the bookends from the first to the last interval of the 8 intervals. The gray area represent where values overlap or crossover.

k – resulting signed-digit {-2, -1, 0, +1, +2}
p – redundancy factor 0.667
j – iterations 1 – 8. Iteration 0 corresponds to initial values.
$\wedge S[j]$ – estimated result from beginning to end is 0.1000 to 0.1111, or 0.5 to 0.9375, an estimate of the partial result corresponding to intervals 0 to 7.

Note: *S[j] instead of A[j] is used here.*

$Uk[j] = 2S[j](k + p) + (k + p)^2 \cdot 4^{-(j+1)}$ Upper boundary

$Lk[j] = 2S[j](k - p) + (k - p)^2 \cdot 4^{-(j+1)}$ Lower boundary

k	Uk [1]	Lk [1]	Uk [8]	Lk [8]
+2	3.111	1.444	5.000	2.499
+1	1.840	0.339	3.125	0.624
0	0.694	-0.639	1.250	-1.250
-1	-0.326	-1.493	-0.624	-3.125
-2	-1.221	-2.222	-2.499	-5.000

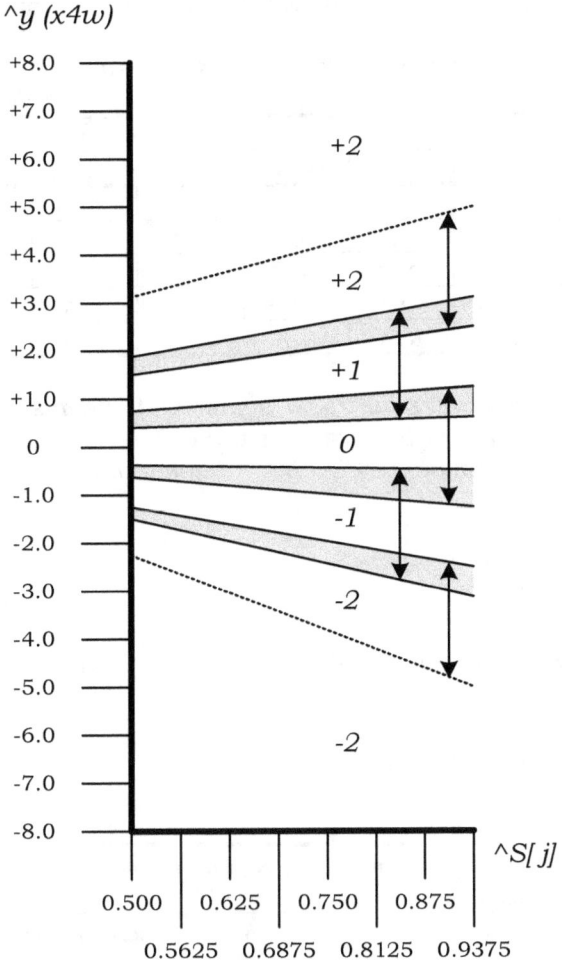

Figure 10

Note: $^\wedge S[j]$ estimate is partitioned into 8 intervals.

Crossover

Crossover in Figure 10 is shown in gray. The next table tabulates the threshold for the signed-digits across the 8 intervals. $^\wedge S[j]$ is the estimate of the partial result, $^\wedge y$ represents the intersecting point for the that interval. Resulting in the selection of S_{j+1}.

Table entries are thresholds for signed-digit values, as expressed below.

$$Sj+1 = k, \text{ threshold } k \le {}^\wedge y < \text{threshold } k + 1$$

k	${}^\wedge S[j]$							
	0.500	0.5625	0.625	0.6875	0.750	0.8125	0.875	0.9375
+2	1.5	1.75	2.0	2.0	2.25	2.5	2.5	2.75
+1	0.5	0.5	0.5	0.5	0.75	0.75	1.0	1.0
0	-0.5	-0.625	-0.75	-0.75	-0.75	-1.0	-1.0	-1.0
-1	-1.625	-1.75	-2.0	-2.125	-2.25	-2.5	-2.75	-2.875
-2	N/A	N/A	N/A	N/A	N/A	N/A	N/A	N/A

Note: *the interval values for the -1 signed-digit seems lower than the boundaries provided in the previous table. These thresholds were published by Ercegovac and Lang, and some may have been arrived at empirically.*

6.2.9 Carry Select Adders

See section 7.2.6 of "STATE MACHINES IN VHDL Dividers Vol. 3".

6.2.10 De-normalization

Internal Operand Size		*ncnt*	Action taken
integer	fractional		
0	>0	0	No action operand fractional
0	>0	>0	Right shift $ncnt/2$
>0	0	N/A	Right shift Integer size – ((integer size – $ncnt$)/2)
>0	>0	0	Right shift Integer size/2
>0	>0	>0	Normalizing was within integer portion Right shift Integer size – ((integer size – ncnt)/2)
>0	>0	>0	Normalizing was within fractional portion Right shift Integer size + (fcnt/2)

Note: *integer and fractional size are those of the internal operand.*

De-normalization converts the resulting fractional root, which is greater or equal to ½ and less than 1, to the initial fixed-point format. Several parameters determine how many right shifts are required to properly locate the data bits relative to the original fixed-point position. Specifically, the internal operand integer and fractional sizes and the recorded shifts during normalization (*ncnt/fcnt*).

The second step is to right shift the internal working register up to 2-bits at a time until the result is properly aligned. As lower bits are shifted they are used to set the sticky bit.

Note: *General rule -- the root of an integer or a fixed-point number with an integer payload (regardless if there is a fractional content) is smaller than the radicand. In contrast, the root of a fractionally only payload is larger.*

6.2.11 Remapping

Remapping takes the de-normalized internal register and properly maps each bit into its proper location using the integer and fractional portions of the input/output fixed-point format.

No lower bits are lost in this process but are accounted for in the pre-rounding step.

6.2.12 Correction, Rounding, and Sticky Bit Decode

As state earlier the result in two's complement is contain in register A upon completion of the algorithm, register B is $A - 1$. The sign of the last residual determines whether the actual result should be A or B. First the carry-save form of the residual, ws and wc, must be added to obtain the signed two's complement equivalent. If the value is positive register A is the result, if negative B is the result. Furthermore, a positive non-zero residual also sets the sticky bit.

Note: *conversion of carry-save to two's complement is done by a multi-cycle carry-select adder.*

Two extra bits guard and round are added to the internal operand registers. These bits along with the LSB of the de-normalized value and the sticky bit are used to determine whether the result is rounded. The rounding function can be enabled or disabled at the start of any root extraction cycle.

Note: *rounding is done by a multi-cycle carry-select adder.*

Sticky is set if the residual is a non-zero positive value, or if any lower 1 bits are shifted out during de-normalization, or if the extra LSB bit that was added to make the fractional portion even is set.

Note: *See chapter 5 for Normalizing, Rounding, and Bounds.*

6.2.13 Conclusion

Performance

The number of clock cycles required to complete the operation is approximately the number of bits in the internal working registers minus the sign bit. Additional overhead includes steps for converting negative operands, mapping/remapping, normalizing/de-normalizing (this is data dependent), and rounding.

Note: *See chapter 5 for Normalizing, Rounding, and Bounds.*

Optional Improvements

1). Full operand lengths were employed here to process on ongoing residual to the full length of the input/output fixed-point.

- Perfect roots only require a limited number of iterations. The best solution is to add circuits that watch for the residual becoming zero then exit the algorithm early.

  In order to maintain the targeted clock speed, values from *ws* and *wc* would have to enter a pipelined adder. This will cause a multi-clock latency but would allow an early result to exit the algorithm sooner.

- Reduce the number of iterations in creating the lower bits. There is no guideline for this, only that it should not be any less than half. This is a typical resolution required for square roots.

- If the input operand data bits only occupy the lower half of the fractional portion, the number of iterations should be INT_SIZ+((FRAC_SIZ/2)). The lower bits would only be shifted out, setting the sticky bit.

2). De-normalization right shifts the fractional result into its former fixed-point position. Increase the number of maximum shifts per cycle to beyond 2. If resources are available a barrel shifter could accomplish this in 1 clock.

Example Design

The example state machine provided illustrates the *Radix-4 Square Root.*

- Format is scalable fixed-point Qm.n, integer and fractional bit lengths as generic parameters, defaulting to Q31.32, with a range to Q63.64. It is recommended that the combined size of integer and fractional portions be 8 or larger.
- Signed two's complement numbers are supported for the input radicand. All conversions are handled and managed internally by the state machine. An imaginary bit is used to represent a negative output.
- Rounding is dynamic, being enabled or disabled during each operation. Defaults to round-to-nearest-even.
- The design has intentional duplication in order to reduce fan out.

As configured for Q31.32, test builds were run using Xilinx ISE.

The following performance was obtained with corresponding parts.

Xilinx Spartan XC3s500e greater than 151MHz
Xilinx Virtex XC5vlx30 greater than 275MHz

```
-----------------------------------------------------
--
--   Radix4Sqrt.vhd
--
-----------------------------------------------------
library IEEE;
use IEEE.std_logic_1164.all;
use IEEE.numeric_std.all;
-- user packages
use work.sd4_pkg.all;


entity Radix4Sqrt is
--
--   Qmn fixed point format is used.
--
--          sign        binary point
--           |             |
--   format <s>(integer bits).(fractional bits)
--           _____/ _____/
--              INT_SIZ       FRAC_SIZ
--
--   Note: full range of operands are supported. The operand length must be
--   even, so the design scales internal registers to accomplish this.
--
--   However, a minimum of 7-bits should be used, otherwise the design may not
--   operate properly. This is a combined number of integer and fractional
--   bits, which will also set the range of operation.
--
generic (INT_SIZ: integer range 0 to 63 := 31;
         FRAC_SIZ: integer range 0 to 64 := 32);
port
(
    clk: in std_logic; -- system clock
    rst: in std_logic;  -- system reset (must be synchronous)
    -- inputs
    rnd_en: in std_logic; -- enable rounding
    extr: in std_logic; -- initiate root extraction
    rcd: in signed((1+INT_SIZ+FRAC_SIZ)-1 downto 0); -- radicand
    -- outputs
    rdy: out std_logic; -- data ready
    udr: out std_logic; -- underflow
    i: out std_logic; -- imaginary
    rt: out signed((1 + INT_SIZ + FRAC_SIZ)-1 downto 0) -- root
```

```
);
end Radix4Sqrt;

architecture RTL of Radix4Sqrt is


--------------
-- functions
--------------
function make_even_siz(x: integer) return integer is
begin

    if(x = 0) then
        return(0);
    -- odd number of bits, so return even
    elsif(x rem 2 /= 0) then
        return(x + 1);
    -- even number of bits, leave as is
    else
        return(x);
    end if;

end function;


-------------------------------
--   type and declared constants
-------------------------------
-- operands must be even
constant OP_INT_SIZ: integer := make_even_siz(INT_SIZ);
constant OP_FRAC_SIZ: integer := make_even_siz(FRAC_SIZ);

-- sign along with guard and round bits included
constant OP_SIZ: integer := 1 + OP_INT_SIZ + OP_FRAC_SIZ + 2;


--------------
-- components
--------------
component rds is
port
(
    clk: in std_logic;
    -- inputs
    j: in integer range 0 to 1;
    \^y\: in unsigned(6 downto 0);
    A0: in std_logic; -- s.1fff
```

```vhdl
        A2: in std_logic;
        A3: in std_logic;
        A4: in std_logic;
        -- output
        sd4: out signed(2 downto 0)
);
end component;


component csla is
-- operand size must be greater or equal to partition size
generic (SIZ,PRTN_SIZ: integer);
port
(
        clk: in std_logic;
        rst: in std_logic;
        -- inputs
        wr: in std_logic;
        ci: in std_logic;
        x: in unsigned(SIZ-1 downto 0);
        y: in unsigned(SIZ-1 downto 0);
        -- outputs
        rdy: out std_logic;
        so: out unsigned(SIZ-1 downto 0);
        co: out std_logic
);
end component;


----------------------
--   declared signals
----------------------
signal A,B,NOT_A: signed(OP_SIZ-1 downto 0) := (others=>'0');
signal new_A_for_sd4_plus_2: signed(OP_SIZ-1 downto 0) := (others=>'0');
signal new_A_for_sd4_plus_1: signed(OP_SIZ-1 downto 0) := (others=>'0');
signal new_A_for_sd4_0: signed(OP_SIZ-1 downto 0) := (others=>'0');
signal new_A_for_sd4_minus_1: signed(OP_SIZ-1 downto 0) := (others=>'0');
signal new_A_for_sd4_minus_2: signed(OP_SIZ-1 downto 0) := (others=>'0');
signal new_B_for_sd4_plus_2: signed(OP_SIZ-1 downto 0) := (others=>'0');
signal new_B_for_sd4_plus_1: signed(OP_SIZ-1 downto 0) := (others=>'0');
signal new_B_for_sd4_0: signed(OP_SIZ-1 downto 0) := (others=>'0');
signal new_B_for_sd4_minus_1: signed(OP_SIZ-1 downto 0) := (others=>'0');
signal new_B_for_sd4_minus_2: signed(OP_SIZ-1 downto 0) := (others=>'0');

signal F: signed((OP_SIZ+2)-1 downto 0) := (others=>'0');
signal prev_A_x2_for_F: signed(F'high downto 0) := (others=>'0');
```

```
signal prev_A_x4_for_F: signed(F'high downto 0) := (others=>'0');
signal prev_B_x2_for_F: signed(F'high downto 0) := (others=>'0');
signal prev_B_x4_for_F: signed(F'high downto 0) := (others=>'0');

signal apnd_A_x4_to_F: signed(F'high downto 0) := (others=>'0');
signal apnd_A_x2_to_F: signed(F'high downto 0) := (others=>'0');
signal apnd_B_x2_to_F: signed(F'high downto 0) := (others=>'0');
signal apnd_B_x4_to_F: signed(F'high downto 0) := (others=>'0');

-- A/B/F lower bit position for concatenation
signal a_lwr_pos: natural range 0 to (OP_SIZ - 1) := 0;
signal b_lwr_pos: natural range 0 to (OP_SIZ - 1) := 0;
signal f_lwr_pos: natural range 0 to (OP_SIZ - 1) := 0;

signal ws: unsigned((OP_SIZ+1)-1 downto 0) := (others=>'0');
signal wc: unsigned((OP_SIZ+1) downto 1) := (others=>'0');

signal x4ws: unsigned((OP_SIZ+3)-1 downto 0) := (others=>'0');
signal x4wc: unsigned((OP_SIZ+3) downto 1) := (others=>'0');

signal x4wsr: unsigned((OP_SIZ+3)-1 downto 0) := (others=>'0');
signal x4wcr: unsigned((OP_SIZ+3) downto 1) := (others=>'0');
signal x4w_est: unsigned(7 downto 0) := (others=>'0');

signal adder_in1: unsigned((OP_SIZ+1)-1 downto 0) := (others=>'0');
signal adder_in2: unsigned((OP_SIZ+1)-1 downto 0) := (others=>'0');
signal adder_out: unsigned((OP_SIZ+1)-1 downto 0) := (others=>'0');

signal rcd_in: signed(rcd'high downto rcd'low) := (others=>'0');

signal tmp1,tmp2,tmp3,tmp4,tmp5: unsigned(OP_SIZ-1 downto 0) := (others=>'0');

signal clsa_rdy,csla_wr: std_logic := '0';

signal npos: natural range 0 to tmp1'high := 0;
signal ncnt: natural range 0 to OP_SIZ := 0;
signal dcnt: natural range 0 to OP_SIZ := 0;
signal fcnt: natural range 0 to OP_SIZ := 0;

signal portion: std_logic := '0';
signal grs: unsigned(2 downto 0) := "000";
signal lsb: std_logic := '0';
signal j: natural range 0 to 1 := 0;
```

```vhdl
signal \S[j+1]_a\: signed(2 downto 0) := (others=>'0');
signal \S[j+1]_b\: signed(2 downto 0) := (others=>'0');
signal \S[j+1]_f\: signed(2 downto 0) := (others=>'0');


--
-- preventing the removal of registers and signals reduces fanout
-- and improves performance.
--
attribute equivalent_register_removal: string;
attribute equivalent_register_removal of A : signal is "no";
attribute equivalent_register_removal of B : signal is "no";
attribute equivalent_register_removal of F : signal is "no";
attribute equivalent_register_removal of a_lwr_pos : signal is "no";
attribute equivalent_register_removal of b_lwr_pos : signal is "no";
attribute equivalent_register_removal of f_lwr_pos : signal is "no";

attribute keep:string;
attribute keep of A :signal is "true";
attribute keep of B :signal is "true";
attribute keep of F :signal is "true";
attribute keep of a_lwr_pos :signal is "true";
attribute keep of b_lwr_pos :signal is "true";
attribute keep of f_lwr_pos :signal is "true";

attribute keep of \S[j+1]_a\ :signal is "true";
attribute keep of \S[j+1]_b\ :signal is "true";
attribute keep of \S[j+1]_f\ :signal is "true";

attribute keep_hierarchy:string;
attribute keep_hierarchy of RTL : architecture is "true";


----------------------
--   enumeration lists
----------------------
type sm_def is
(
    RESET,
    START_SQRT,
    MAP_BITS,
    CNVRT_TO_POS,
    CNVRT_TO_POS2,
    NORMALIZE_OP,
    NORMALIZE_OP2,
    PREP_FOR_EXTR,
```

```
    SQRT_ALGO,
    SQRT_ALGO2,
    REM_COR,
    REM_COR2,
    REM_COR3A,
    REM_COR3B,
    REM_COR3C,
    DENORMALIZE,
    DENORMALIZE2,
    PREROUND,
    ROUND,
    ROUND2,
    COMPLETE,
    ERROR
);
signal state: sm_def := RESET;

---------------------------------- module code ----------------------------
begin

NOT_A <= not A;
----------------------------------------
--
--   Radix-4 Square Root state machine
--
----------------------------------------
process(rst,clk)
begin

    if(rst='1') then

        -- working registers
        A <= (others=>'0');
        B <= (others=>'0');
        F <= (others=>'0');
        tmp1 <= (others=>'0');
        tmp2 <= (others=>'0');
        tmp3 <= (others=>'0');
        tmp4 <= (others=>'0');
        tmp5 <= (others=>'0');
        x4wsr <= (others=>'0');
        x4wcr <= (others=>'0');

        adder_in1 <= (others=>'0');
```

```
adder_in2  <= (others=>'0');

prev_A_x4_for_F <= (others=>'0');
prev_A_x2_for_F <= (others=>'0');
prev_B_x2_for_F <= (others=>'0');
prev_B_x4_for_F <= (others=>'0');
apnd_A_x4_to_F <= (others=>'0');
apnd_A_x2_to_F <= (others=>'0');
apnd_B_x2_to_F <= (others=>'0');
apnd_B_x4_to_F <= (others=>'0');

new_A_for_sd4_plus_2 <= (others=>'0');
new_A_for_sd4_plus_1 <= (others=>'0');
new_A_for_sd4_0 <= (others=>'0');
new_A_for_sd4_minus_1 <= (others=>'0');
new_A_for_sd4_minus_2 <= (others=>'0');

new_B_for_sd4_plus_2 <= (others=>'0');
new_B_for_sd4_plus_1 <= (others=>'0');
new_B_for_sd4_0 <= (others=>'0');
new_B_for_sd4_minus_1 <= (others=>'0');
new_B_for_sd4_minus_2 <= (others=>'0');

-- misc
a_lwr_pos <= 0;
b_lwr_pos <= 0;
f_lwr_pos <= 0;
portion <= '0';
j <= 0;

-- control
csla_wr <= '0';
npos <= 0;
ncnt <= 0;
dcnt <= 0;
fcnt <= 0;
lsb <= '0';

-- handshake and error signals
udr <= '0';
i <= '0';
rt <= (others=>'0');

-- states
```

```vhdl
        state <= RESET;

elsif rising_edge(clk) then

    -- one clock signals
    csla_wr <= '0';


    --
    --   state machine body
    --
    case state is
        -- reset state
        when RESET =>
            state <= START_SQRT;
        --
        --   square root body
        --
        when START_SQRT =>
            -- wait for extraction command
            if(extr = '1') then
                udr <= '0';
                rcd_in <= rcd;
                ncnt <= 0;
                fcnt <= 0;
                portion <= '0';
                state <= MAP_BITS;
            end if;
            -- point to first fractional bits
            npos <= tmp1'high - 2;
        --
        -- map radicand into local operand
        --
        when MAP_BITS =>
            -- map fractional bits
            if(FRAC_SIZ /= 0) then
                -- when equal just include gard and round
                if(OP_FRAC_SIZ = FRAC_SIZ) then
                    tmp1((FRAC_SIZ+2)-1 downto 0) <=
                        unsigned(rcd_in(FRAC_SIZ-1 downto 0))&"00";
                -- or larger by 1 account for extra bit
                else
                    tmp1((FRAC_SIZ+3)-1 downto 0) <=
                        unsigned(rcd_in(FRAC_SIZ-1 downto 0))&"000";
                end if;
```

```
        end if;
        -- map sign and integer bits
        if(INT_SIZ /= 0) then
            -- when equal
            if(OP_INT_SIZ = INT_SIZ) then
                tmp1(tmp1'high downto OP_FRAC_SIZ+2) <=
                    unsigned(rcd_in(rcd_in'high downto FRAC_SIZ));
            -- or larger by 1 sign extend
            else
                tmp1(tmp1'high-1 downto OP_FRAC_SIZ+2) <=


                    unsigned(rcd_in(rcd_in'high downto FRAC_SIZ));
                tmp1(tmp1'high) <= rcd_in(rcd_in'high);
            end if;
        -- account for sign bit only
        else
            tmp1(tmp1'high) <= rcd_in(rcd_in'high);
        end if;
        -- convert to positive if negative
        if(rcd_in(rcd_in'high) = '1') then
            i <= '1';
            state <= CNVRT_TO_POS;
        else
            i <= '0';
            state <= NORMALIZE_OP;
        end if;
    --
    -- convert radicand to positive operand using fast adder.
    --
    when CNVRT_TO_POS =>
        adder_in1 <= not(tmp1(tmp1'high)&tmp1);
        adder_in2 <= to_unsigned(1,adder_in2'length);
        csla_wr <= '1';
        state <= CNVRT_TO_POS2;
    when CNVRT_TO_POS2 =>
        if(clsa_rdy = '1') then
            tmp2 <= adder_out(adder_out'high-1 downto 0);
            state <= NORMALIZE_OP2;
        end if;
    --
    -- normalize operand to between 1/4 and 3/4 (< 1) by left
    -- shifting 2-bits at a time. if the content is zero its an
    -- error. ignore sign bit, it should always be zero.
```

```
--
when NORMALIZE_OP =>
    tmp2 <= tmp1;
    state <= NORMALIZE_OP2;
when NORMALIZE_OP2 =>
    -- locate leading digit
    if(tmp2(tmp2'high - 1 downto tmp2'high - 2) /= "00") then
            tmp3 <= tmp2;
            state <= PREP_FOR_EXTR;
    -- check for lower boundary of operand
    elsif(npos > 0) then
        npos <= npos - 2;
        ncnt <= ncnt + 2;
        tmp2(tmp2'high-1 downto 0) <=
            tmp2(tmp2'high - 3 downto 0)&"00";
    -- lower boundary reached, operand is zero
    else
        state <= ERROR;
    end if;
    -- determine portion of operand where leading digit are
    if(ncnt = OP_INT_SIZ) then
        portion <= '1';
    elsif(ncnt > OP_INT_SIZ) then
        fcnt <= fcnt + 2;
    end if;
--
-- prepare variables for extraction
--
when PREP_FOR_EXTR =>
    A <= (others=>'0');
    A(A'high) <= '1'; -- for p < 1
    B <= (others=>'0');
    F <= (others=>'0');
    a_lwr_pos <= A'high - 2;
    b_lwr_pos <= B'high - 2;
    f_lwr_pos <= F'high - 4;
    -- add -1 to radicand, but no shift, for first S[j+1]

    x4wsr <= "1111"&tmp3(tmp3'high-1 downto 0);
    x4wcr <= (others=>'0');
    j <= 0;
    grs <= "000";
    state <= SQRT_ALGO;
--
```

```
--  square root algorithm
--

when SQRT_ALGO =>
    --
    -- provide anticipated value for F from both A and B
    --
    -- Sj+1 = +2
    prev_A_x4_for_F <= (others=>'0');
    if(j = 0) then
        prev_A_x4_for_F(F'high) <= NOT_A(A'high);
    else
        prev_A_x4_for_F(F'high downto f_lwr_pos+4) <=
            NOT_A(A'high downto f_lwr_pos+2);
    end if;
    -- Sj+1 = +1
    prev_A_x2_for_F <= (others=>'0');
    if(j = 0)then
        prev_A_x2_for_F(F'high) <= '1';
        prev_A_x2_for_F(F'high-1) <= NOT_A(A'high);
    else
        prev_A_x2_for_F(F'high) <= '1';
        prev_A_x2_for_F(F'high-1 downto f_lwr_pos+3) <=
            NOT_A(A'high downto f_lwr_pos+2);
    end if;
    -- Sj+1 = -1
    prev_B_x2_for_F <= (others=>'0');
    if(j = 0) then
        prev_B_x2_for_F(F'high) <= '0';
        prev_B_x2_for_F(F'high-1) <= B(B'high);
    else
        prev_B_x2_for_F(F'high-1 downto f_lwr_pos+3) <=
            B(B'high downto f_lwr_pos+2);
    end if;
    -- Sj+1 = -2
    prev_B_x4_for_F <= (others=>'0');
    if(j = 0) then
        prev_B_x4_for_F(F'high) <= B(B'high);
    else
        prev_B_x4_for_F(F'high downto f_lwr_pos+4) <=
            B(B'high downto f_lwr_pos+2);
    end if;


    --
    -- anticipated value for F's appendage to A and B
```

```
  --
  apnd_A_x4_to_F <= (others=>'0');
  apnd_A_x4_to_F(f_lwr_pos+3 downto f_lwr_pos) <= "1100";
  apnd_A_x2_to_F <= (others=>'0');
  apnd_A_x2_to_F(f_lwr_pos+2 downto f_lwr_pos) <= "111";
  apnd_B_x2_to_F <= (others=>'0');
  apnd_B_x2_to_F(f_lwr_pos+2 downto f_lwr_pos) <= "111";
  apnd_B_x4_to_F <= (others=>'0');
  apnd_B_x4_to_F(f_lwr_pos+3 downto f_lwr_pos) <= "1100";


  --
  -- anticipated values for registers A and B
  --
  new_A_for_sd4_plus_2  <= A;
  new_A_for_sd4_plus_2(a_lwr_pos+1 downto a_lwr_pos) <= "10";
  new_A_for_sd4_plus_1  <= A;
  new_A_for_sd4_plus_1(a_lwr_pos+1 downto a_lwr_pos) <= "01";
  new_A_for_sd4_0 <= A;
  new_A_for_sd4_0(a_lwr_pos+1 downto a_lwr_pos) <= "00";
  new_A_for_sd4_minus_1    <= B;
  new_A_for_sd4_minus_1(a_lwr_pos+1 downto a_lwr_pos) <= "11";
  new_A_for_sd4_minus_2 <= B;
  new_A_for_sd4_minus_2(a_lwr_pos+1 downto a_lwr_pos) <= "10";


  new_B_for_sd4_plus_2  <= A;
  new_B_for_sd4_plus_2(b_lwr_pos+1 downto b_lwr_pos) <= "01";
  new_B_for_sd4_plus_1  <= A;
  new_B_for_sd4_plus_1(b_lwr_pos+1 downto b_lwr_pos) <= "00";
  new_B_for_sd4_0 <= B;
  new_B_for_sd4_0(b_lwr_pos+1 downto b_lwr_pos) <= "11";
  new_B_for_sd4_minus_1 <= B;
  new_B_for_sd4_minus_1(b_lwr_pos+1 downto b_lwr_pos) <= "10";
  new_B_for_sd4_minus_2    <= B;
  new_B_for_sd4_minus_2(b_lwr_pos+1 downto b_lwr_pos) <= "01";

  state <= SQRT_ALGO2;

when SQRT_ALGO2 =>
  j <= 1;-- only 0 and 1 have any bearing
  --
  -- update registers A and B based on S[j+1]
  --
  -- update register A
  case \S[j+1]_a\ is
```

```
        when \SD4 +2\ =>    A <= new_A_for_sd4_plus_2;
        when \SD4 +1\ =>    A <= new_A_for_sd4_plus_1;
        when \SD4 -1\ =>    A <= new_A_for_sd4_minus_1;
        when \SD4 -2\ =>    A <= new_A_for_sd4_minus_2;
        when others     => A <= new_A_for_sd4_0; -- \SD4  0\
    end case;
    -- update register B
    case \S[j+1]_b\ is
        when \SD4 +2\ =>    B <= new_B_for_sd4_plus_2;
        when \SD4 +1\ =>    B <= new_B_for_sd4_plus_1;
        when \SD4 -1\ =>    B <= new_B_for_sd4_minus_1;
        when \SD4 -2\ =>    B <= new_B_for_sd4_minus_2;
        when others     => B <= new_B_for_sd4_0; -- \SD4  0\
    end case;


    --
    -- update register F  based on S[j+1]
    -- (uses previously computed values)
    --
    case \S[j+1]_f\ is
        when \SD4 +2\ =>    F <= prev_A_x4_for_F or apnd_A_x4_to_F
        when \SD4 +1\ =>    F <= prev_A_x2_for_F or apnd_A_x2_to_F;
        when \SD4 -1\ =>    F <= prev_B_x2_for_F or apnd_B_x2_to_F;
        when \SD4 -2\ =>    F <= prev_B_x4_for_F or apnd_B_x4_to_F;
        when others =>      F <= (others=>'0');    -- \SD4  0\
    end case;

    -- subsequent S[j+1]
    x4wsr  <= x4ws;
    x4wcr <= x4wc;

    -- update register lower offset for next cycle, exit if last
    -- iteration (b and f are identical)
    if(a_lwr_pos = 0) then
        state <= REM_COR;
    else
        a_lwr_pos <= a_lwr_pos - 2;
        b_lwr_pos <= b_lwr_pos - 2;
        f_lwr_pos <= f_lwr_pos - 2;
        state <= SQRT_ALGO;
    end if;


    --
    --  Remainder and root correction/conversion
```

```vhdl
--
when REM_COR =>
    -- convert carry-save to two's complement
    csla_wr <= '1';
    adder_in1 <= ws(ws'high downto 0);
    adder_in2 <= wc(wc'high-1 downto 1)&'0';
    state <= REM_COR2;
when REM_COR2 =>
    if(clsa_rdy = '1') then
        -- check to see if residual is negative
        if(adder_out(adder_out'high) = '1') then
            state <= REM_COR3C;
        -- residual zero
        elsif(adder_out = 0) then
            state <= REM_COR3A;
        -- residual positive
        else
            state <= REM_COR3B;
        end if;
    end if;
-- the residual was zero
when REM_COR3A =>
    tmp4 <= unsigned(A);
    grs(0) <= '0';
    state <= DENORMALIZE;
-- the residual was positive
when REM_COR3B =>
    tmp4 <= unsigned(A);
    grs(0) <= '1';
    state <= DENORMALIZE;
-- the residual was negative (too far)
when REM_COR3C =>
    tmp4 <= unsigned(B);
    grs(0) <= '0';
    state <= DENORMALIZE;
--
--  Final steps
--
when DENORMALIZE =>
    -- fractionally only
    if(OP_INT_SIZ = 0 and OP_FRAC_SIZ > 0) then
        -- fractionally aligned
        if(ncnt = 0) then
            dcnt <= 0;
```

```vhdl
        -- fractionally unaligned
        else
            dcnt <= (ncnt/2);
        end if;
    -- integer only
    elsif(OP_INT_SIZ > 0 and OP_FRAC_SIZ = 0) then
        dcnt <= OP_INT_SIZ - (OP_INT_SIZ - ncnt)/2;
    -- integer and fractional portions
    else
        -- upper integer bits set
        if(ncnt = 0) then
            dcnt <= OP_INT_SIZ/2;
        -- full range of values
        else
            -- within integer portion
            if(portion = '0') then
                dcnt <= OP_INT_SIZ - (OP_INT_SIZ - ncnt)/2;
            -- within fractional portion
            else
                dcnt <= OP_INT_SIZ + (fcnt/2);
            end if;
        end if;
    end if;
    state <= DENORMALIZE2;
-- do it fast
when DENORMALIZE2 =>
    if(dcnt = 0) then
        state <= PREROUND;
    elsif(dcnt = 1) then
        dcnt <= dcnt - 1;
      tmp4(tmp4'high) <= '0';
        tmp4(tmp4'high-1 downto 0) <= "0"&tmp4(tmp4'high-1 downto 1);
        -- set sticky if any lower true bits are discarded
        grs(0) <= grs(0) or tmp4(0);
        state <= PREROUND;
    elsif(dcnt = 2) then
        dcnt <= dcnt - 2;
        tmp4(tmp4'high) <= '0';
        tmp4(tmp4'high-1 downto 0) <= "00"&tmp4(tmp4'high-1 downto
2);
        -- set sticky if any lower true bits are discarded
        grs(0) <= grs(0) or tmp4(1) or tmp4(0);

        state <= PREROUND;
```

```
        else -- dcnt > 2
            dcnt <= dcnt - 2;
            tmp4(tmp4'high) <= '0';
            tmp4(tmp4'high-1 downto 0) <=
                "00"&tmp4(tmp4'high-1 downto 2);
            -- set sticky if any lower true bits are discarded
            grs(0) <= grs(0) or tmp4(1) or tmp4(0);
        end if;
    when PREROUND =>
        -- prepare operand for rounding, extract guard and round
        if(FRAC_SIZ /= 0) then
            -- no extra bit
            if(OP_FRAC_SIZ = FRAC_SIZ) then
                tmp5 <= tmp4(tmp4'high downto 2)&"00";
                grs(2 downto 1) <= tmp4(1 downto 0);
                lsb <= tmp4(2);
              -- or extra bit with sticky
            else
                tmp5 <= tmp4(tmp4'high downto 3)&"000";
                grs(2 downto 1) <= tmp4(2 downto 1);
                grs(0) <= grs(0) or tmp4(0);
                lsb <= tmp4(3);
            end if;
        -- no fractional portion
        else
                tmp5 <= tmp4(tmp4'high downto 2)&"00";
                grs(2 downto 1) <= tmp4(1 downto 0);
                lsb <= tmp4(2);
        end if;
        state <= ROUND;
    when ROUND =>
        -- check if rounding is enabled, then do so based on rules
        if(rnd_en = '1' and(grs > 4 or (grs = 4 and lsb = '1'))) then
            adder_in1 <= tmp5(tmp5'high)&tmp5;
            -- account for guard and round offset
            if(FRAC_SIZ /= 0) then
                    -- offset different with extra bit
                    if(OP_FRAC_SIZ = FRAC_SIZ) then

                        adder_in2 <= to_unsigned(4,adder_in2'length);
                    else
                        adder_in2 <= to_unsigned(8,adder_in2'length);
                    end if;
            -- no fractional portion
```

```
        else
            adder_in2 <= to_unsigned(4,adder_in2'length);
        end if;
        csla_wr <= '1';
        state <= ROUND2;
    else
        state <= COMPLETE;
    end if;
when ROUND2 =>
    if(clsa_rdy = '1') then
        tmp5 <= adder_out(adder_out'high-1 downto 0);
        state <= COMPLETE;
    end if;
-- remap bits to output
when COMPLETE =>
    -- map fractional bits
    if(FRAC_SIZ /= 0) then
        -- transfer over, minus rounding bits
        if(OP_FRAC_SIZ = FRAC_SIZ) then
            rt(FRAC_SIZ-1 downto 0) <=
                signed(tmp5((FRAC_SIZ+2)-1 downto 2));
        -- or larger by 1 account for extra bit
        else
            rt(FRAC_SIZ-1 downto 0) <=
                signed(tmp5((FRAC_SIZ+3)-1 downto 3));
        end if;
    end if;
    -- map sign and integer bits
    if(INT_SIZ /= 0) then
        -- include two rounding bits
        if(OP_FRAC_SIZ = FRAC_SIZ) then
            rt(INT_SIZ+FRAC_SIZ-1 downto FRAC_SIZ) <=
        signed(tmp5((INT_SIZ+FRAC_SIZ+2)-1 downto FRAC_SIZ+2));
        -- include extra bit
        else
            rt(INT_SIZ+FRAC_SIZ-1 downto FRAC_SIZ) <=
        signed(tmp5((INT_SIZ+FRAC_SIZ+3)-1 downto FRAC_SIZ+3));
        end if;
    end if;
    -- sign is always positive
    rt(rt'high) <= '0';
    state <= START_SQRT;
--
-- error occurred
```

```
            --
         when ERROR =>
             udr <= '1';
             state <= START_SQRT;

         when others => state <= RESET;
       end case;
     end if;

end process;

rdy <= '1' when (extr = '0') and (state = START_SQRT) else '0';


    --
    --  carry save adder to compute new residual w[j+1] in
    --  carry-save form. only the upper two bits of the new
    --  residual are used in the next cycle.
    --
gen_csa: for i in 0 to ws'high generate
begin

    gen_lsb: if i = 0 generate
        ws(i) <= x4wsr(i) xor F(i);
        wc(i+1) <= x4wsr(i) and F(i); -- cin is zero
    end generate;

    gen_other: if i > 0 generate
        ws(i) <= (x4wsr(i) xor x4wcr(i)) xor F(i);
        wc(i+1) <= (x4wsr(i) and F(i)) or (x4wcr(i) and F(i)) or (x4wsr(i) and
x4wcr(i));
    end generate;

end generate;

-- multiply w (residual) times 4
x4ws <= ws&"00";
x4wc <= wc&"00";

--  carry propagation adder for table input, iiii.ffff (4 integer, 4 fractional)
x4w_est <=
    x4ws(x4ws'high downto x4ws'high-7) + x4wc(x4wc'high-1 downto x4wc'high-8);


    --
```

```vhdl
--   Result Digit Selection table (duplication reduces fanout of sd4)
--
rds_mod_a: rds
port map
(
    clk => clk,
    -- inputs
    j => j,
    \^y\ => x4w_est(7 downto 1),-- iiii.fff
    A0 => A(A'high), -- s.1fff
    A2 => A(A'high-2),
    A3 => A(A'high-3),
    A4 => A(A'high-4),
    -- output
    sd4 => \S[j+1]_a\
);

rds_mod_b: rds
port map
(
    clk => clk,
    -- inputs
    j => j,
    \^y\ =>  x4w_est(7 downto 1),-- iiii.fff
    A0 => A(A'high), -- s.1fff
    A2 => A(A'high-2),
    A3 => A(A'high-3),
    A4 => A(A'high-4),
    -- output
    sd4 => \S[j+1]_b\
);

rds_mod_f: rds
port map
(
    clk => clk,
    -- inputs
    j => j,
    \^y\ =>  x4w_est(7 downto 1),-- iiii.fff
    A0 => A(A'high), -- s.1fff
    A2 => A(A'high-2),
    A3 => A(A'high-3),
    A4 => A(A'high-4),
    -- output
```

```
        sd4 => \S[j+1]_f\
);


--
--   General use Carry Select Adder (fast adder)
--
csla_mod: csla
generic map (SIZ => OP_SIZ+1, PRTN_SIZ => 8)
port map
(
        clk => clk,
        rst => rst,
        -- inputs
        wr => csla_wr,
        ci => '0',
        x => adder_in1,
        y => adder_in2,
        -- outputs
        rdy => clsa_rdy,
        so => adder_out,
        co => open
);


end RTL;
```

```
--------------------------------------------------------------------------------
--
--   SD4_pkg.vhd (Signed Digit radix 4 package file)
--
--
--------------------------------------------------------------------------------
library IEEE;
use IEEE.std_logic_1164.all;
use IEEE.numeric_std.all;


--
--  package header
--
package SD4_pkg is

-- radix 4 signed digit constants
constant \SD4 +2\: signed(2 downto 0) := "010";
constant \SD4 +1\: signed(2 downto 0) := "001";
constant \SD4  0\: signed(2 downto 0) := "000";
constant \SD4 -1\: signed(2 downto 0) := "111";
constant \SD4 -2\: signed(2 downto 0) := "110";


--
-- rom table type
--
--   Design notes: in order for incoming address into array to have a range from
--   +/- 8.000, the ROM table must be arranged properly. The first half of the
--   table corresponds to an input spanning the positive range, the second the
--   negative in two's complement.
--
--   0 to 511 positive index, 512 down to 1023 negative index (top down).
--
--   The sign, integer bits, and most significant three fractional bits of the
--   address are indexed  by residual estimate, or /^y/. The lowest three fractional
--   bits are indexed by the most significant partial result bits, or /^S/, and
--   from left to right.
--
type sd4_array is array (natural range <>) of integer range -2 to +2; --signed(2
downto 0);
constant ResultDigitTable: sd4_array(0 to 1023) := -- 10-bit address
(
-- from left to right value, lower three digits (xxx)
--
--0.5000    0.5625 0.6250 0.6875 0.7500 0.8125 0.8750 0.9375 Decimal
```

```
--0.1000    0.1001 0.1010 0.1011 0.1100 0.1101 0.1110 0.1111 Binary
--
--                                              actual        Decimal
--                                              10-bit
--                                              address
    0, 0, 0, 0, 0, 0, 0, 0, -- 0000.000xxx +0.000
    0, 0, 0, 0, 0, 0, 0, 0, -- 0000.001xxx +0.125
    0, 0, 0, 0, 0, 0, 0, 0, -- 0000.010xxx +0.250
    0, 0, 0, 0, 0, 0, 0, 0, -- 0000.011xxx +0.375
    +1,+1,+1,+1, 0, 0, 0, 0, -- 0000.100xxx  +0.500
    +1,+1,+1,+1, 0, 0, 0, 0, -- 0000.101xxx  +0.625
    +1,+1,+1,+1,+1,+1, 0, 0, -- 0000.110xxx    +0.750
    +1,+1,+1,+1,+1,+1, 0, 0, -- 0000.111xxx    +0.875

    +1,+1,+1,+1,+1,+1,+1,+1, -- 0001.000xxx    +1.000
    +1,+1,+1,+1,+1,+1,+1,+1, -- 0001.001xxx    +1.125
    +1,+1,+1,+1,+1,+1,+1,+1, -- 0001.010xxx    +1.250
    +1,+1,+1,+1,+1,+1,+1,+1, -- 0001.011xxx    +1.375
    +2,+1,+1,+1,+1,+1,+1,+1, -- 0001.100xxx    +1.500
    +2,+1,+1,+1,+1,+1,+1,+1, -- 0001.101xxx    +1.625
    +2,+2,+1,+1,+1,+1,+1,+1, -- 0001.110xxx    +1.750
    +2,+2,+1,+1,+1,+1,+1,+1, -- 0001.111xxx    +1.875

    +2,+2,+2,+2,+1,+1,+1,+1, -- 0010.000xxx    +2.000
    +2,+2,+2,+2,+1,+1,+1,+1, -- 0010.001xxx    +2.125
    +2,+2,+2,+2,+2,+1,+1,+1, -- 0010.010xxx    +2.250
    +2,+2,+2,+2,+2,+1,+1,+1, -- 0010.011xxx    +2.375
    +2,+2,+2,+2,+2,+2,+2,+1, -- 0010.100xxx    +2.500
    +2,+2,+2,+2,+2,+2,+2,+1, -- 0010.101xxx    +2.625
    +2,+2,+2,+2,+2,+2,+2, -- 0010.110xxx    +2.750
    +2,+2,+2,+2,+2,+2,+2, -- 0010.111xxx    +2.875

    +2,+2,+2,+2,+2,+2,+2,+2, -- 0011.000xxx    +3.000
    +2,+2,+2,+2,+2,+2,+2,+2, -- 0011.001xxx    +3.125
    +2,+2,+2,+2,+2,+2,+2,+2, -- 0011.010xxx    +3.250
    +2,+2,+2,+2,+2,+2,+2,+2, -- 0011.011xxx    +3.375
    +2,+2,+2,+2,+2,+2,+2,+2, -- 0011.100xxx    +3.500
    +2,+2,+2,+2,+2,+2,+2,+2, -- 0011.101xxx    +3.625
    +2,+2,+2,+2,+2,+2,+2,+2, -- 0011.110xxx    +3.750
    +2,+2,+2,+2,+2,+2,+2,+2, -- 0011.111xxx    +3.875

    +2,+2,+2,+2,+2,+2,+2,+2, -- 0100.000xxx    +4.000
    +2,+2,+2,+2,+2,+2,+2,+2, -- 0100.001xxx    +4.125
    +2,+2,+2,+2,+2,+2,+2,+2, -- 0100.010xxx    +4.250
```

69

```
+2,+2,+2,+2,+2,+2,+2,+2, -- 0100.011xxx   +4.375
+2,+2,+2,+2,+2,+2,+2,+2, -- 0100.100xxx   +4.500
+2,+2,+2,+2,+2,+2,+2,+2, -- 0100.101xxx   +4.625
+2,+2,+2,+2,+2,+2,+2,+2, -- 0100.110xxx   +4.750
+2,+2,+2,+2,+2,+2,+2,+2, -- 0100.111xxx   +4.875

+2,+2,+2,+2,+2,+2,+2,+2, -- 0101.000xxx   +5.000
+2,+2,+2,+2,+2,+2,+2,+2, -- 0101.001xxx   +5.125
+2,+2,+2,+2,+2,+2,+2,+2, -- 0101.010xxx   +5.250
+2,+2,+2,+2,+2,+2,+2,+2, -- 0101.011xxx   +5.375
+2,+2,+2,+2,+2,+2,+2,+2, -- 0101.100xxx   +5.500
+2,+2,+2,+2,+2,+2,+2,+2, -- 0101.101xxx   +5.625
+2,+2,+2,+2,+2,+2,+2,+2, -- 0101.110xxx   +5.750
+2,+2,+2,+2,+2,+2,+2,+2, -- 0101.111xxx   +5.875

+2,+2,+2,+2,+2,+2,+2,+2, -- 0110.000xxx   +6.000
+2,+2,+2,+2,+2,+2,+2,+2, -- 0110.001xxx   +6.125
+2,+2,+2,+2,+2,+2,+2,+2, -- 0110.010xxx   +6.250
+2,+2,+2,+2,+2,+2,+2,+2, -- 0110.011xxx   +6.375
+2,+2,+2,+2,+2,+2,+2,+2, -- 0110.100xxx   +6.500
+2,+2,+2,+2,+2,+2,+2,+2, -- 0110.101xxx   +6.625
+2,+2,+2,+2,+2,+2,+2,+2, -- 0110.110xxx   +6.750
+2,+2,+2,+2,+2,+2,+2,+2, -- 0110.111xxx   +6.875

+2,+2,+2,+2,+2,+2,+2,+2, -- 0111.000xxx   +7.000
+2,+2,+2,+2,+2,+2,+2,+2, -- 0111.001xxx   +7.125
+2,+2,+2,+2,+2,+2,+2,+2, -- 0111.010xxx   +7.250
+2,+2,+2,+2,+2,+2,+2,+2, -- 0111.011xxx   +7.375
+2,+2,+2,+2,+2,+2,+2,+2, -- 0111.100xxx   +7.500
+2,+2,+2,+2,+2,+2,+2,+2, -- 0111.101xxx   +7.625
+2,+2,+2,+2,+2,+2,+2,+2, -- 0111.110xxx   +7.750
+2,+2,+2,+2,+2,+2,+2,+2, -- 0111.111xxx   +7.875
-- boundary between positive and negative
   -2,-2,-2,-2,-2,-2,-2,-2, -- 1000.000xxx   -8.000

   -2,-2,-2,-2,-2,-2,-2,-2, -- 1000.001xxx   -7.875
   -2,-2,-2,-2,-2,-2,-2,-2, -- 1000.010xxx   -7.750
   -2,-2,-2,-2,-2,-2,-2,-2, -- 1000.011xxx   -7.625
   -2,-2,-2,-2,-2,-2,-2,-2, -- 1000.100xxx   -7.500
   -2,-2,-2,-2,-2,-2,-2,-2, -- 1000.101xxx   -7.375
   -2,-2,-2,-2,-2,-2,-2,-2, -- 1000.110xxx   -7.250
   -2,-2,-2,-2,-2,-2,-2,-2, -- 1000.111xxx   -7.125
   -2,-2,-2,-2,-2,-2,-2,-2, -- 1001.000xxx   -7.000
```

```
-2,-2,-2,-2,-2,-2,-2,-2, -- 1001.001xxx    -6.875
-2,-2,-2,-2,-2,-2,-2,-2, -- 1001.010xxx    -6.750
-2,-2,-2,-2,-2,-2,-2,-2, -- 1001.011xxx    -6.625
-2,-2,-2,-2,-2,-2,-2,-2, -- 1001.100xxx    -6.500
-2,-2,-2,-2,-2,-2,-2,-2, -- 1001.101xxx    -6.375
-2,-2,-2,-2,-2,-2,-2,-2, -- 1001.110xxx    -6.250
-2,-2,-2,-2,-2,-2,-2,-2, -- 1001.111xxx    -6.125
-2,-2,-2,-2,-2,-2,-2,-2, -- 1010.000xxx    -6.000

-2,-2,-2,-2,-2,-2,-2,-2, -- 1010.001xxx    -5.875
-2,-2,-2,-2,-2,-2,-2,-2, -- 1010.010xxx    -5.750
-2,-2,-2,-2,-2,-2,-2,-2, -- 1010.011xxx    -5.625
-2,-2,-2,-2,-2,-2,-2,-2, -- 1010.100xxx    -5.500
-2,-2,-2,-2,-2,-2,-2,-2, -- 1010.101xxx    -5.375
-2,-2,-2,-2,-2,-2,-2,-2, -- 1010.110xxx    -5.250
-2,-2,-2,-2,-2,-2,-2,-2, -- 1010.111xxx    -5.125
-2,-2,-2,-2,-2,-2,-2,-2, -- 1011.000xxx    -5.000

-2,-2,-2,-2,-2,-2,-2,-2, -- 1011.001xxx    -4.875
-2,-2,-2,-2,-2,-2,-2,-2, -- 1011.010xxx    -4.750
-2,-2,-2,-2,-2,-2,-2,-2, -- 1011.011xxx    -4.625
-2,-2,-2,-2,-2,-2,-2,-2, -- 1011.100xxx    -4.500
-2,-2,-2,-2,-2,-2,-2,-2, -- 1011.101xxx    -4.375
-2,-2,-2,-2,-2,-2,-2,-2, -- 1011.110xxx    -4.250
-2,-2,-2,-2,-2,-2,-2,-2, -- 1011.111xxx    -4.125
-2,-2,-2,-2,-2,-2,-2,-2, -- 1100.000xxx    -4.000

-2,-2,-2,-2,-2,-2,-2,-2, -- 1100.001xxx    -3.875
-2,-2,-2,-2,-2,-2,-2,-2, -- 1100.010xxx    -3.750
-2,-2,-2,-2,-2,-2,-2,-2, -- 1100.011xxx    -3.625
-2,-2,-2,-2,-2,-2,-2,-2, -- 1100.100xxx    -3.500
-2,-2,-2,-2,-2,-2,-2,-2, -- 1100.101xxx    -3.375
-2,-2,-2,-2,-2,-2,-2,-2, -- 1100.110xxx    -3.250
-2,-2,-2,-2,-2,-2,-2,-2, -- 1100.111xxx    -3.125
-2,-2,-2,-2,-2,-2,-2,-2, -- 1101.000xxx    -3.000

-2,-2,-2,-2,-2,-2,-2,-1, -- 1101.001xxx    -2.875
-2,-2,-2,-2,-2,-2,-1,-1, -- 1101.010xxx    -2.750
-2,-2,-2,-2,-2,-2,-1,-1, -- 1101.011xxx    -2.625
-2,-2,-2,-2,-2,-1,-1,-1, -- 1101.100xxx    -2.500
-2,-2,-2,-2,-2,-1,-1,-1, -- 1101.101xxx    -2.375
-2,-2,-2,-2,-1,-1,-1,-1, -- 1101.110xxx    -2.250
-2,-2,-2,-1,-1,-1,-1,-1, -- 1101.111xxx    -2.125
-2,-2,-1,-1,-1,-1,-1,-1, -- 1110.000xxx    -2.000
```

```
    -2,-2,-1,-1,-1,-1,-1,-1, -- 1110.001xxx    -1.875
    -2,-1,-1,-1,-1,-1,-1,-1, -- 1110.010xxx    -1.750
    -1,-1,-1,-1,-1,-1,-1,-1, -- 1110.011xxx    -1.625
    -1,-1,-1,-1,-1,-1,-1,-1, -- 1110.100xxx    -1.500
    -1,-1,-1,-1,-1,-1,-1,-1, -- 1110.101xxx    -1.375
    -1,-1,-1,-1,-1,-1,-1,-1, -- 1110.110xxx    -1.250
    -1,-1,-1,-1,-1,-1,-1,-1, -- 1110.111xxx    -1.125
    -1,-1,-1,-1,-1, 0, 0, 0, -- 1111.000xxx    -1.000

    -1,-1,-1,-1,-1, 0, 0, 0, -- 1111.001xxx    -0.875
    -1,-1, 0, 0, 0, 0, 0, 0, -- 1111.010xxx -0.750
    -1, 0, 0, 0, 0, 0, 0, 0, -- 1111.011xxx -0.625
     0, 0, 0, 0, 0, 0, 0, 0, -- 1111.100xxx -0.500
     0, 0, 0, 0, 0, 0, 0, 0, -- 1111.101xxx -0.375
     0, 0, 0, 0, 0, 0, 0, 0, -- 1111.110xxx -0.250
     0, 0, 0, 0, 0, 0, 0, 0  -- 1111.111xxx -0.125
);


end;
--
--  package body
--
package body SD4_pkg is


end ;
```

```
--------------------------------------------------------------------------
--
--    rds.vhd (result digit selection table)
--
--    A 10-bit input to a lookup table produces a 3-bit signed digit ranging from
--    -2 to +2. The upper 7-bits are the result of converting the upper 8-bits
--    of the residual in carry-saved-form, to twos' complement. The lower 3-bits
--    of the address are derived from the upper 5-bits of the partial result.
--
--    In order to map memory contiguously over a range over a +/- memory address
--    the sign bit keeps its polarity while all lower bits, excluding the lowest
--    3-bits, are inverted.
--------------------------------------------------------------------------
library IEEE;
use IEEE.std_logic_1164.all;
use IEEE.numeric_std.all;
-- user packages
use work.sd4_pkg.all;

entity rds is
port
(
    clk: in std_logic; -- system clock
    -- inputs
    j: in integer range 0 to 1;
    \^y\: in unsigned(6 downto 0);-- iiii.fff of 4w[j] (residual estimate)
    A0: in std_logic; -- s.1fff
    A2: in std_logic;
    A3: in std_logic;
    A4: in std_logic;
    -- output
    sd4: out signed(2 downto 0) -- signed digit, radix 4 output
);
end rds;

architecture RTL of rds is

signal adr_map: unsigned(9 downto 0) := (others=>'0');
signal adr: integer range 0 to 1023 := 0;
signal \^S\: unsigned(2 to 4) := (others=>'0');



begin
```

-- convert partial result bits to estimate, which also represent
-- the lower address bits into the table.
\^S\(2) <= A0 or A2;
\^S\(3) <= '1' when A3 = '1' or (A0 = '1' and j > 0) else '0';
\^S\(4) <= A0 or A4;

-- map actual address bits based on sign for proper ROM table indexing
adr_map <= \^y\(6)&(\^y\(5 downto 0))&\^S\(2 to 4);

-- convert unsigned to integer for indexing
adr <= to_integer('0'&adr_map);

-- synchronous ROM (inferred READ-FIRST)
process(clk)
begin

    if rising_edge(clk) then
        sd4 <= to_signed(ResultDigitTable(adr),3);
    end if;

end process;

end RTL;

```
-----------------------------------------------------------
--
--   csla.vhd Carry Select Adder module
--
-- Multi-stage carry select adder. Number of stages is based
-- on the operand size and the number of partitions needed
-- to support the operand size.
-----------------------------------------------------------
library IEEE;
use IEEE.std_logic_1164.all;
use IEEE.numeric_std.all;

entity csla is
-- operand size must be greater or equal to partition size
generic (SIZ,PRTN_SIZ: integer);
port
(
    clk: in std_logic; -- system clock
    rst: in std_logic;  -- system reset (must be synchronous)
    -- inputs
    wr: in std_logic;
    ci: in std_logic;
    x: in unsigned(SIZ-1 downto 0);
    y: in unsigned(SIZ-1 downto 0);
    -- outputs
    rdy: out std_logic;
    so: out unsigned(SIZ-1 downto 0);
    co: out std_logic
);
end csla;

architecture RTL of csla is

-- functions
function gen_total_seg_reg(extra,base: integer) return integer is
variable cnt: integer := base;
begin
    if(extra > 0) then
        cnt := cnt + 1;
    end if;
    return(cnt);
end function;

-- data segmenting
```

```vhdl
type seg_type is array (integer range <>) of unsigned(PRTN_SIZ-1 downto 0);

-- sizing constants
constant EXTRA_BITS: integer := SIZ rem PRTN_SIZ;
constant NUM_SEG_REGS: integer := SIZ / PRTN_SIZ;
constant TOTAL_SEG_REG: integer :=
gen_total_seg_reg(EXTRA_BITS,NUM_SEG_REGS);

-- local signals
signal a,b,s: unsigned(x'length-1 downto 0) := (others=>'0');
signal c,c_reg: unsigned(TOTAL_SEG_REG-1 downto 0) := (others=>'0');
signal rdy_dly: unsigned(TOTAL_SEG_REG-1 downto 0) := (others=>'0');

-- test signals
signal s0,s1: seg_type(1 to NUM_SEG_REGS-1) := (others=>(others=>'0'));
signal c0,c1: unsigned(1 to NUM_SEG_REGS-1) := (others=>'0');

----------------------------------- module code -----------------------------
begin

-- register process
process(rst,clk)
begin

    if(rst='1') then
        a <= (others=>'0');
        b <= (others=>'0');
        c_reg <= (others=>'0');
        so <= (others=>'0');
        rdy_dly <= (others=>'1');
    elsif rising_edge(clk) then
        -- input operands
        if(wr = '1') then
            a <= x;
            b <= y;
        end if;
        -- register selected sums and carries
        c_reg <= c;
        so <= s;
        -- ready delay logic
        if(wr = '1') then
            rdy_dly <= (others=>'0');
        else
            rdy_dly <= rdy_dly(rdy_dly'high-1 downto 0)&'1';
```

```
      end if;
    end if;

end process;
-- last carry in chain is output
co <= c_reg(c_reg'high);
rdy <= rdy_dly(rdy_dly'high) when wr = '0' else '0';

-- base segment adder
base_seg:entity work.cpa generic map(PRTN_SIZ)
port map
(   ci, -- carry in from parent module
    a(PRTN_SIZ-1 downto 0),
    b(PRTN_SIZ-1 downto 0),
    s(PRTN_SIZ-1 downto 0),
    c(0)
);

-- adders for upper segments (minus base segment)
gen_upr_seg: if NUM_SEG_REGS > 1 generate
--signal s0,s1: seg_type(1 to NUM_SEG_REGS-1) := (others=>(others=>'0'));
--signal c0,c1: unsigned(1 to NUM_SEG_REGS-1) := (others=>'0');
begin

    -- create adders and interconnecting muxes
    gen_upr_adr: for i in 1 to NUM_SEG_REGS-1 generate
        -- compute segment for each carry input
        upr_cpa_0:entity work.cpa generic map(PRTN_SIZ)
        port map
        (   '0',-- carry in is = 0
            a((PRTN_SIZ*i)+PRTN_SIZ-1 downto (PRTN_SIZ*i)),
            b((PRTN_SIZ*i)+PRTN_SIZ-1 downto (PRTN_SIZ*i)),
            s0(i),c0(i)
        );
        upr_cpa_1:entity work.cpa generic map(PRTN_SIZ)
        port map
        (   '1',-- carry in is = 1
            a((PRTN_SIZ*i)+PRTN_SIZ-1 downto (PRTN_SIZ*i)),
            b((PRTN_SIZ*i)+PRTN_SIZ-1 downto (PRTN_SIZ*i)),
            s1(i),c1(i)
        );
        -- select sum and carries for current segment, based on previous carry
        c(i) <= c0(i) when c_reg(i-1) = '0' else c1(i);
```

```
      s((PRTN_SIZ*i)+PRTN_SIZ-1 downto (PRTN_SIZ*i)) <= s0(i) when c_reg(i-
1) = '0' else s1(i);
    end generate;

end generate;

-- extra-bits adder
gen_xtr_bits: if EXTRA_BITS > 0 generate
signal s0,s1: unsigned(EXTRA_BITS-1 downto 0) := (others=>'0');
signal c0,c1: std_logic := '0';
begin

    xtr_cpa_0:entity work.cpa generic map(EXTRA_BITS)
    port map
    (   '0',-- carry in is = 0
        a(a'high downto (PRTN_SIZ*NUM_SEG_REGS)),
        b(b'high downto (PRTN_SIZ*NUM_SEG_REGS)),
        s0,c0
    );
    xtr_cpa_1:entity work.cpa generic map(EXTRA_BITS)
    port map
    (   '1',-- carry in is = 1
        a(a'high downto (PRTN_SIZ*NUM_SEG_REGS)),
        b(b'high downto (PRTN_SIZ*NUM_SEG_REGS)),
        s1,c1
    );
    -- select sum and carries for current segment
    c(c'high) <= c0 when c_reg(c_reg'high-1) = '0' else c1;
        s(s'high downto s'high-(EXTRA_BITS-1)) <= s0 when c_reg(c_reg'high-1) =
            '0' else s1;

end generate;

end RTL;
```

```
----------------------------------------------------------
--
--   cpa.vhd Carry Propagation Adder
--
----------------------------------------------------------
library IEEE;
use IEEE.std_logic_1164.all;
use IEEE.numeric_std.all;

entity cpa is
generic (SIZ: integer);
port
(
    -- inputs
    ci: in std_logic;
    a: in unsigned(SIZ-1 downto 0);
    b: in unsigned(SIZ-1 downto 0);
    -- outputs
    so: out unsigned(SIZ-1 downto 0);
    co: out std_logic
);
end cpa;

architecture RTL of cpa is

signal c: unsigned(SIZ-1 downto 0) := (others=>'0');

begin

-- created with full adders
gen_adr: for i in 0 to SIZ-1 generate

    gen_lsb: if i = 0 generate
        so(i) <= a(i) xor b(i) xor ci;
        c(i) <= (a(i) and ci) or (b(i) and ci) or (a(i) and b(i));
    end generate;

    gen_other: if i > 0 generate
        so(i) <= a(i) xor b(i) xor c(i-1);
        c(i) <= (a(i) and c(i-1)) or (b(i) and c(i-1)) or (a(i) and b(i));
    end generate;

end generate;
```

```
co <= c(c'high);

end RTL;
```

7 Cube Root Functions

Cube root ($\sqrt[3]{x}$) functions are second in importance as compared to square roots. The ratio between the radicand and root is 3 to 1, instead of 2 to 1. Meaning 1-bit is retired for every 3-bits examined in the radicand. While this is a very simplified comparison, cube root functions require more steps and considerably more resources. Because of this last point two designs are provided. The first design is simple and efficient (efficient for cube roots) and the second high performance (high performance for cube roots). The latter implementation is more appropriate for high-end FPGA devices.

The simple design uses iterative shifts while adding and subtracting terms of partial products; whereas the high performance design uses digit recurrence and is very similar to the *Radix-4 Square Root* design in section 6.2. However, several high-end functions set this design apart like multi-level carry-save adder trees to reduce propagation delays as well as a 2-digit by a full operand multiplier (in carry-save form), and barrel shifters that act as signed-digit by full operand multipliers.

> Note: *Cube roots do not require an imaginary bit because it is possible to have either a positive or negative cube root. All cubes are computed as positive operands, so negative inputs are converted to positive operands and where required the result is converted to negative.*

7.1 Simple Cube Root

Even a simple cube root function is more difficult to design, but attempts were made to maintain a modest level of complexity. Example, only carry propagation adders are used. Multiplication is done by adding shifted partial products. Excessive delay buildup from multiple levels of propagation adders is alleviated by creative pipelining. As a result, only 2 clock cycles are required to retire each extracted root bit. As with other designs in this book the operand and result are fixed-point and scalable and have parity between input and output in terms of their respective Qm.n formats.

The algorithm employed here was referenced by Hong Peng [2] and is fixed-point, meaning that no normalization is required. However the incoming operand must be composed in 3-bit groups relative to the binary point. The design pads both the integer and fraction portions automatically. These additional bits are added between the MSB and the sign and additional bits are added to the right of the LSB, respectively. Figure 11.

Note: *Variable names and terms with the original algorithm are maintained here.*

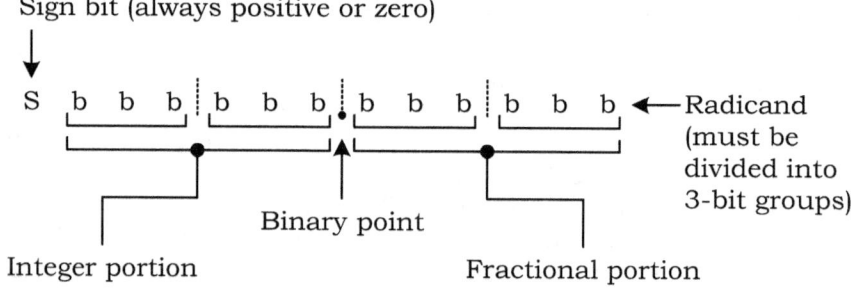

Figure 11

Working Registers

There are four primary registers. Each are explained below

Register A – Is the working radicand that contains the original operand value but with zero padded 3-bit grouping. During extraction the upper 3-bits, excluding the sign, are evaluated then left shifted 3-bits for the next cycle. These upper 3-bits are represented by J (excluding sign) which is an alias name.

Register R – Contains the working root as bits are retired. As a bit is retired register R is left shifted 1-bit and the new bit is placed in the LSB position. The content of register R is fed back into the algorithm throughout the extraction. The length of R also includes space for the extra guard and round bits for rounding.

Register C – Contains the remainder following each cycle and is fed back into the algorithm.

Register P – It called the determinant. It contains the combined elements of registers R and the previous value of P.

Note: *The bracket [ ] associated with each register signifies the register value at that iteration.*

Steps in Basic Algorithm

The actual algorithm is explained in the next section because the required pipelining make it hard to recognize. Here the steps are simplified.

1). Initialization

$i = 0$
A = operand (J alias to A(A'high-1 downto A'high-3))
$R = 0$
$C = 0$
$P = 1$

2). Set first bit

$R = 1$ if $(J - P) \geq 0$
$C = J - P$
$R = 0$ else
$C = 0$

$A = 8A$
$i = 1$

3). Loop

Create determinant for current cycle based on LSB of $R[i]$ and P
$P[i] = 4P[i-1] + 18R[i] - 3$ if LSB of $R[i] = 1$
$P[i] = 4P[i-1] - 6R[i] - 3$ else

Compare remainder to determinant then update C and R
$C[i+1] = (8C[i] + Ji+1) - P[i]$ if expression ≥ 0
$R[i+1] = 2R[i] + 1$
$C[i+1] = (8C[i] + Ji+1)$ else
$R[i+1] = 2R[i] + 0$

Position next 3-bits of working radicand
$A = 8A$
$i = i+1$

The number of Iterations is equal to the number of combined integer and fraction portions of the input/output Qm.n format plus rounding bits.

4). The result R must be right shifted to re-establish the original binary point.

7.1.1 Architecture Overview

Figure 12 represents the *Simple Cube Root* design. The design is scalable supporting a full range of Qm.n formats, combined integer and fractional portions or either integer or fractional only. The input radicand is mapped into an internal working register while adding pad bits to both the integer and factional portions to create the 3-bit grouping required by the algorithm. This is done automatically through instantiation by way of the integer and fractional size generics. A negative input must first be converted to positive, which is required by algorithm.

Since the algorithm is for fixed-point operands no normalizing or shifting is required, extraction can begin immediately.

Three states accomplish the extraction: Set First Bit, COMPUTE PI, and DETERMINE RI. The latter two being repeated the number of cycles required to retire all bits involved in the extraction and are heavily pipelined.

Once completed, the result is right shifted to restore the working root register to its proper binary point position. Rounding is done if enabled and the extracted root is available in the original Qm.n format. If required, the root is first converted into a negative number.

Explanation of Variables

Because of the heavy pipelining the expressions are broken up into their individual or paired terms. Additionally, variable i is used slightly different here from the original algorithm. It coincides with the retirement of each root bit not necessarily how the terms and factors are expressed. For example i, $i+1$, and $i-1$ represent the current register value, the next value, and the previous value respectively.

Pipelined registers

$6R$ - contains 6 times the current R value, $6R[i]$.
$18R$ - contains 18 times the current R value, $18R[i]$.
$4P$ - 3 - contains 4 times the previous P value minus 3, $4P[i-1]$ - 3.
$P\_for\_R0\_1$ – contains the pre-computed expression of the
determinant P for the current cycle if the last R bit retired was a 1.

$$P[i] = 4P[i-1] + 18R[i] - 3$$

This expression is constructed with pipelined registers $4P$ - 3
and $18R$.

$P\_for\_R0\_0$ – contains the pre-computed expression of the
determinant P for the current cycle if the last R bit retired was a 0.

$$P[i] = 4P[i-1] - 6R[i] - 3$$

This expression is constructed with pipelined registers $4P$ - 3
and $6R$.

$8C + J$ – contains the pre-computed value of the current remainder
and the next upper 3-bits of the working radicand register A, $8C[i]$
+ $J[i+1]$

Set First Bit (state)

The algorithm is primed by evaluating the first 3-bits, J, of the
working root A to see if it is non-zero. If not zero $R = 1$, $6R = 6$, and
$18R$ is = 18. The remainder C is set to $J - 1$.

Either way P is set to 1, then $4P$ - 3 is set to 1 because it is equal to
$4P$ - 3. Register A is then shifted left 3-bits for next evaluation.

COMPUTE PI (state)

Figure 13 diagrams the pre-computing of pipelined values
$P\_for\_R0\_1$ and $P\_for\_R0\_0$ for the next state, which represents the
choices for the determinant P. Both values are computed so that
the next state can simply select between the two based on the LSB
of R, the last root bit retired.

Additionally $8C_i + J_{i+1}$ is pre-computed by shifting C left 3-bits and adding the current J, indicated by J_{i+1}. This is also used in the next state.

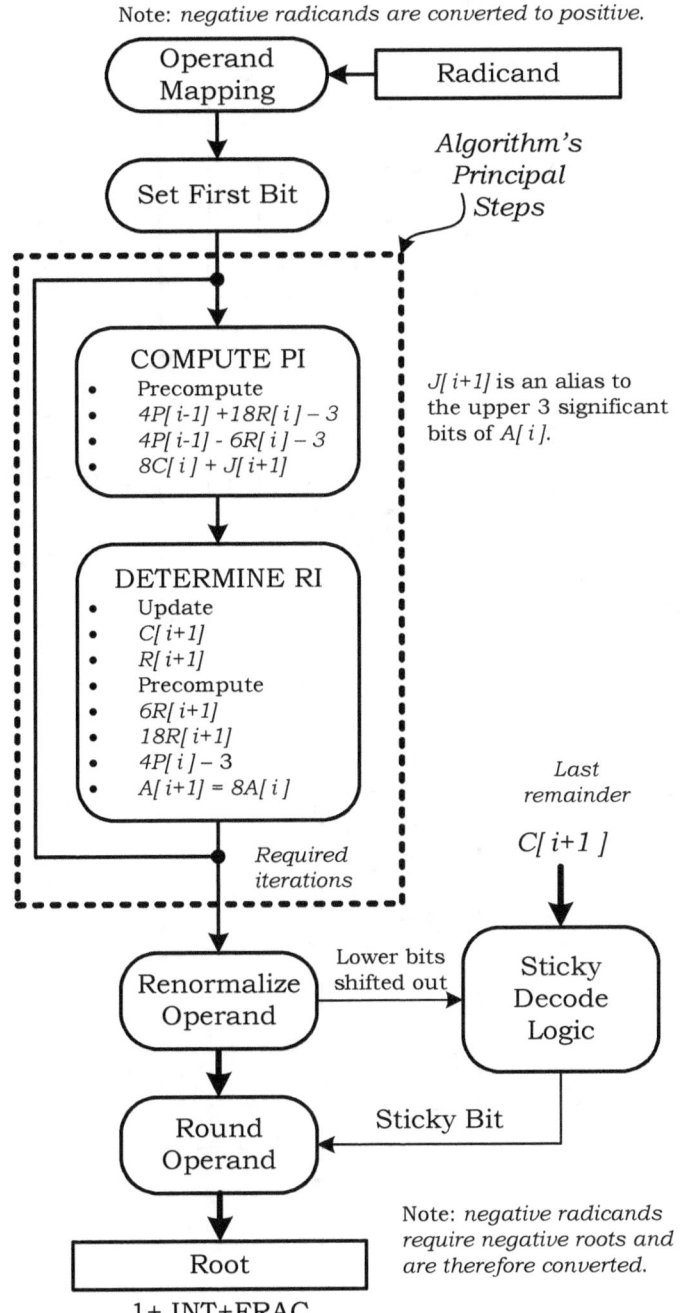

Figure 12

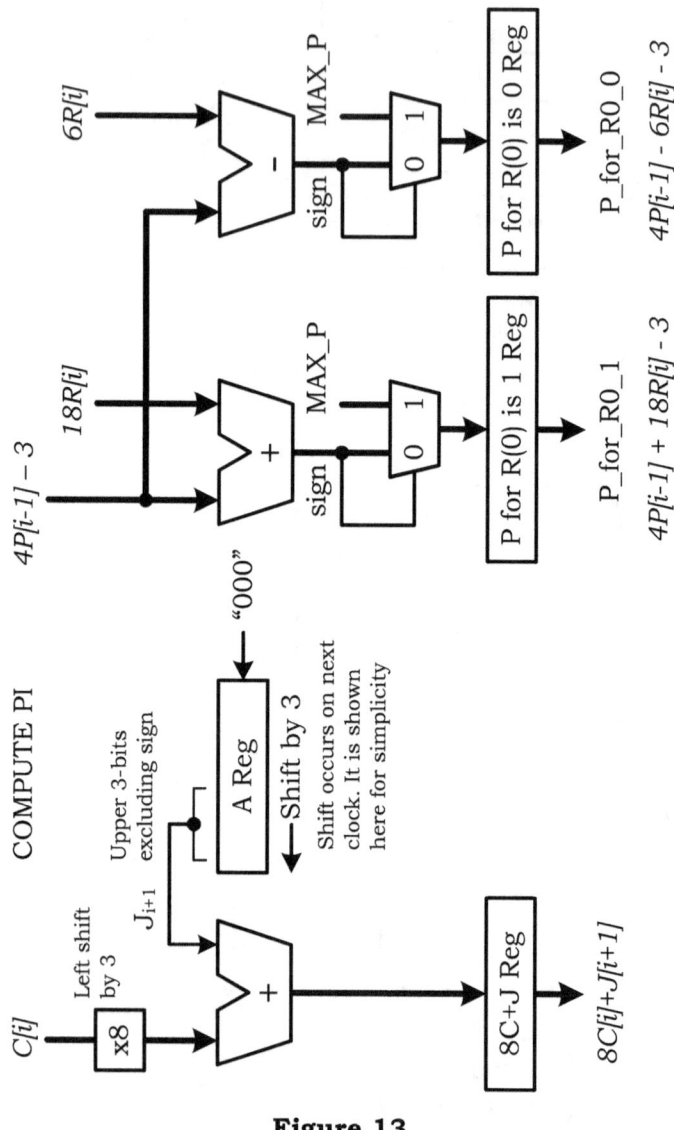

Figure 13

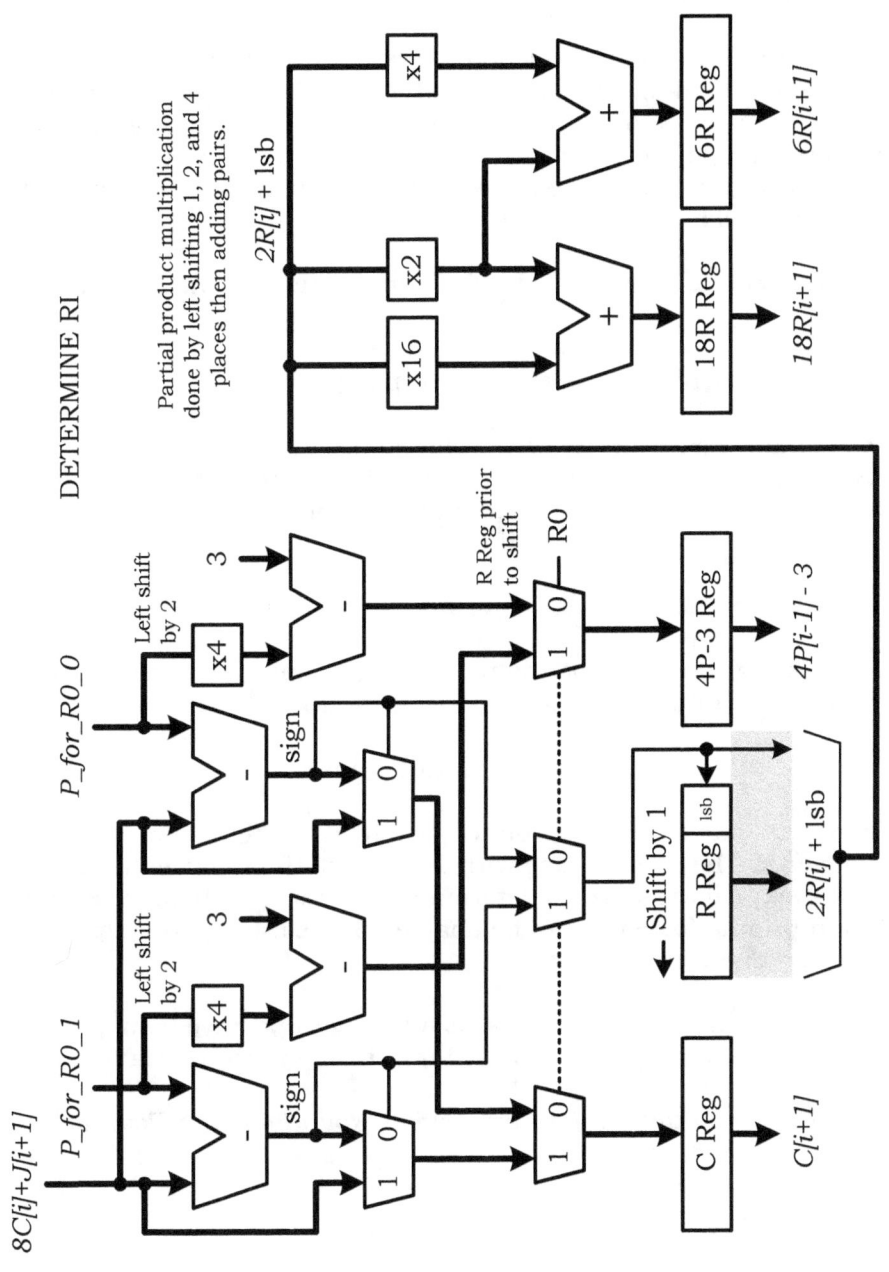

Figure 14

DETERMINE RI (state)

Figure 14 diagrams the selection between the two pre-computed determinant values *P_for_R0_1* and *P_for_R0_1,* and corresponding *8Ci + Ji+1* values for setting of the new root bit, updating the remainder *C*, pre computing the next *4P - 3*, as well as the next *18R* and *16R* value. These are all used in creating the determinant *P* for the next cycle.

The working radicand is left shifted to position the next 3-bits.

7.1.2 Renormalizing and Remapping

Renormalizing in this design is simple because the algorithm is for fixed-point numbers. Re-establishing the leading bit is the same for a given Qm.n format and is not data dependent. The expression for the number of final right shifts is as follows:

Input operand int size – (working radicand int size/3)

Remapping simply maps the resulting data into the output format while maintaining the relative binary point.

7.1.3 Correction, Rounding, and Sticky

Two extra bits guard and round are added to the working root register *R*. These bits along with the LSB of the renormalized value and sticky bit are used to determine whether the result is rounded. The rounding function can be enabled or disabled at the start of any root extraction cycle.

Sticky is set if the remainder *C* is a non-zero positive value, or when any lower 1 bits are shifted out during renormalization.

Note: *See chapter 5 for Normalizing, Rounding, and Bounds.*

7.1.4 Conclusion

Design Limitations

The nature of this algorithm accumulates large values in the *P* register. Primarily because the result *R* continues to increase in magnitude every iteration due to left shifting. Unattended the *P* register will wrap thus halting the creation of lower bits.

A generic parameter was added allowing the primary registers to have extended bits to contain larger numbers. The settings below were empirically arrived at:

Q7.0,	Q0.8	Q7.8	XBITS = 4
Q15.0	Q0.16	Q15.16	XBITS = 12
Q31.0	Q0.32	Q31.32	XBITS = 14 (default)
Q63.0	Q0.64	Q63.64	XBITS = 32

Note(1): *With the input at its highest value, worse case, the lowest 3-bits and 6-bits for Q31.32 and Q63.64 respectively, will be zero in the result. This compromise allows the clock speed to be higher. For better resolution XBITS can be increased 2-3 bits per bit in the result.*

Performance

Each extracted bit (cycle or iteration) takes two clocks. The operand length is the input/output Qm.n size plus 2-bits for guard and round. Overhead includes converting negative operands to positive ones to support the algorithm and back to a negative root if required, one clock each. Two extra clocks are required for the rounding path whether enabled or not. Renormalizing depends on the Qm.n format selected and is not data dependent, taking one clock per shift.

Optional Improvements

1). Full operand lengths were employed here to process ongoing remainders to the full length of the input/output fixed-point format.

- Perfect roots only require a limited number of iterations. The best solution is to add circuits that watch for the remainder and working radicand becoming zero then exit the algorithm early.

- Reduce the number of iterations in creating the lower bits. There are no guidelines for this, only that it should not be any less than one third. This is a typical resolution required for cube roots.

2). Utilize combined carry-save adders along with carry propagation adders instead of carry propagation adders alone.

3). Use a single clock-barrel shifter for renormalizing the result. The shift value is the same for each Qm.n format. It is not data dependent.

4). If rounding is not used remove those two states from design.

Example Design

The example state machine provided illustrates the *Simple Cube Root*.

- Format is scalable fixed-point Qm.n, integer and fractional bit lengths as generic parameters, defaulting to Q31.32, with a range to Q63.64. It is recommend that the combined size of integer and fractional portions be 7 or larger.
- An additional parameter call XBITS allows extending internal adder range for greater number size. Source code contains recommended values for each Qm.n format. Default 14.
- Signed two's complement numbers are supported for the input radicand. All conversions are handled and managed internally by the state machine. A cube root can have a negative result. No imaginary bit is needed.
- Rounding is dynamic, being enabled or disabled during each operation. Defaults to round-to-nearest-even.

As configured for Q31.32, test builds were run using Xilinx ISE.

The following performance was obtained with corresponding parts.

Xilinx Spartan XC3s500e greater than 90MHz
Xilinx Virtex XC5vlx30 greater than 200MHz

```
--------------------------------------------------------------------
--   SimpleCbrt.vhd (Simple Cube root)
--------------------------------------------------------------------
library IEEE;
use IEEE.std_logic_1164.all;
use IEEE.numeric_std.all;
-- user packages
-- use work.ConversionPackageV2.all;

entity SimpleCbrt is
--
--   Qmn fixed point format is used.
--
--         sign        binary point
--          |           |
--   format <s>(integer bits).(fractional bits)
--            _____/ _____/
--            INT_SIZ        FRAC_SIZ
--
--   Note: full range of operands are supported. The operand length must be
--   3-bit groups for both integer portion and fractional portions (minus sign).
--   The design scales internal registers to accomplish this.
--
--   However, a minimum of 7-bits should be used, otherwise the design may not
--   operate properly. This is a combined number of integer and fractional
--   bits, which will also set the range of operation.
--
--   XBITS represent the addition bits need in the primary registers which allow
--   internal adders to grow beyond the base operand size, thus having more
--   precision in the result. Larger values can result in slower clock speeds.
--   Recommended value for standard Qm.n formats.
--
--   Q7.0,     Q0.8      Q7.8      XBITS = 4
--   Q15.0     Q0.16     Q15.16    XBITS = 12
--   Q31.0     Q0.32     Q31.32    XBITS = 14 Note(1) Default
--   Q63.0     Q0.64     Q63.64    XBITS = 32 Note(1)
--
--   Note(1): With the input at its highest value, worse case, the lowest
--   3-bits and 6-bits for Q31.32 and Q63.64 respectively, will be zero
--   in the result. This compromise allows the clock speed to be higher.
--
--   For better resolution XBITS cab be increased 2-3 bits per bit in the
--   result.
--
```

```
generic (INT_SIZ: integer range 0 to 63 := 31;
    FRAC_SIZ: integer range 0 to 64 := 32;
    XBITS: integer range 0 to 32 := 14);
port
(
    clk,rst: in std_logic; -- system clock and reset (synchronous)
    -- inputs
    rnd_en: in std_logic; -- enable rounding for current cycle
    extr: in std_logic; -- initiate root extraction
    rcd: in signed((1+INT_SIZ+FRAC_SIZ)-1 downto 0); -- radicand
    -- outputs
    rdy: out std_logic; -- data ready
    udr: out std_logic; -- underflow
    rt: out signed((1+INT_SIZ+FRAC_SIZ)-1 downto 0) -- root
);
end SimpleCbrt;

architecture RTL of SimpleCbrt is
-----------------------
-- General Functions
-----------------------
function make_groups_of_3_siz(x: integer) return integer is
begin

    if(x = 0) then
        return(0);
    -- if not in groups of 3-bits, adjust
    elsif(x rem 3 = 1) then
        return(x + 2);
    elsif(x rem 3 = 2) then
        return(x + 1);
    -- bits are in groups of 3 already, leave as is
    else
        return(x);
    end if;

end function;
-- add partial products to make times 6
function x6(x: signed; siz: integer) return signed is
variable y: signed(siz-1 downto 0) := (others=>'0');
begin

    y(x'high downto 0) := x;
    return(((y sll 2) + (y sll 1)));
```

```
end function;
-- add partial products to make times 18
function x18(x: signed; siz: integer) return signed is
variable y: signed(siz-1 downto 0) := (others=>'0');
begin

    y(x'high downto 0) := x;
    return(((y sll 4) + (y sll 1)));

end function;


-----------------------
-- declared constants
-----------------------

-- working radicand size, pad to 3-bit boundaries (sign not factored in)
constant WRCD_INT_SIZ: integer := make_groups_of_3_siz(INT_SIZ);
constant WRCD_FRAC_SIZ: integer := make_groups_of_3_siz(FRAC_SIZ);
constant WRCD_SIZ: integer := 1+WRCD_INT_SIZ+WRCD_FRAC_SIZ;

constant RENORMALIZE_CNT: integer := (INT_SIZ)-((WRCD_INT_SIZ)/3);

-- The ratio of significant bits between the working radicand
-- and the working root register is 3:1. When creating the
-- working radicand constants (above), integer and fractional
-- are adjusted to guarantee 3-bit groups. However, to maintain
-- resolution of the input operand the working root register has
-- parity with the root output plus 2 rounding bits.

-- working root are the same size as output except added rounding bits
constant WRT_SIZ: integer := 1+INT_SIZ+FRAC_SIZ+2;

---------------------
-- declared signals
---------------------
signal A: signed(WRCD_SIZ-1 downto 0) := (others=>'0');
alias \J\ is A(A'high-1 downto A'high - 3);
signal sign: std_logic := '0';

signal R: signed(WRT_SIZ-1 downto 0) := (others=>'0');
alias \2R\ is R(R'high-1 downto 0);
signal \6R\: signed((WRT_SIZ+XBITS)-1 downto 0) := (others=>'0');
signal \18R\: signed((WRT_SIZ+XBITS)-1 downto 0) := (others=>'0');
signal \4P-3\: signed((WRT_SIZ+XBITS)-1 downto 0) := (others=>'0');
signal C: signed((WRT_SIZ+XBITS)-1  downto 0) := (others=>'0');
```

```vhdl
signal \8C&J\: signed((WRT_SIZ+XBITS)-1 downto 0) := (others=>'0');

signal P_for_R0_1: signed(\4P-3\'high downto 0) := (others=>'0');
signal P_for_R0_0: signed(\4P-3\'high downto 0) := (others=>'0');

constant ONES_FOR_P: signed(\4P-3\'high-1 downto 0) := (others=>'1');
constant MAX_P: signed(\4P-3\'high downto 0) := '0'&ONES_FOR_P;

signal i: integer range 0 to (R'length-1) := 0;
signal cnt: integer range 0 to RENORMALIZE_CNT := 0;
signal nr: signed((WRT_SIZ-2)-1 downto 0) := (others=>'0');

signal grs: unsigned(2 downto 0) := (others=>'0');

signal busy: std_logic := '0';
signal rnd: std_logic := '0';


-----------------------
--   enumeration lists
-----------------------
type sm_def is
(
    RESET,
    START_EXTR,
    PROCESS_SIGN,
    FIRST_BIT,
    COMPUTE_PI,
    DETERMINE_RI,
    RENORMALIZE,
    ROUND,
    ROUND2,
    CONVRT
);
signal state: sm_def := RESET;

---------------------------------- module code ----------------------------


begin
--------------------------------------
--   Simple Square Root state machine
--------------------------------------
process(rst,clk)
begin
    if(rst='1') then
```

```
    -- inputs
    A <= (others=>'0');
    sign <= '0';
    -- local
    i <= 0;
    R <= (others=>'0');
    \6R\ <= (others=>'0');
    \18R\ <= (others=>'0');
    \4P-3\ <= (others=>'0');
    C <= (others=>'0');
    \8C&J\ <= (others=>'0');
    P_for_R0_1 <= (others=>'0');
    P_for_R0_0 <= (others=>'0');
    cnt <= 0;
    nr <= (others=>'0');
    grs <= (others=>'0');

    -- outputs
    busy<='0'; udr<='0'; rnd <= '0';
    rt <= (others=>'0');
    -- states
    state <= RESET;
elsif rising_edge(clk) then
    --
    --   state machine body
    --
    case state is
        -- reset state
        when RESET =>
            state <= START_EXTR;
        -- wait for signal to start extraction
        when START_EXTR =>
            if(extr = '1') then
                busy<='1'; udr<='0'; rnd <= rnd_en;
                -- account for padding in integer portion by sign extending
                A(A'high downto (A'high - WRCD_INT_SIZ)) <=
        resize(rcd(rcd'high downto (rcd'high - INT_SIZ)),WRCD_INT_SIZ+1);
                -- account for padding in fractional by appending zeros to content
                if(FRAC_SIZ > 0) then
                    A(WRCD_FRAC_SIZ-1 downto 0) <= (others=>'0');
                    A(WRCD_FRAC_SIZ-1 downto WRCD_FRAC_SIZ-
                        FRAC_SIZ) <= rcd(FRAC_SIZ-1 downto 0);
                end if;
                state <= PROCESS_SIGN;
```

```
        end if;
-- convert working register if negative
when PROCESS_SIGN =>
    if(A(A'high) = '1') then
        sign <= '1'; -- save sign
        A <= (not A) + 1;
    end if;
    -- prime algorithm
    i <= 1;
    R <= (others=>'0');
    \6R\ <= (others=>'0');
    \18R\ <= (others=>'0');
    C <= (others=>'0');
    \4P-3\ <= (others=>'0');
    P_for_R0_1 <= (others=>'0');
    P_for_R0_0 <= (others=>'0');
    \8C&J\ <= (others=>'0');
    cnt <= 0;
    nr  <= (others=>'0');
    grs <= "000";
    state <= FIRST_BIT;

--------------------------
--
-- extract root algorithm
--
--------------------------
-- evaluate first bit
when FIRST_BIT =>
    -- set first bit (P0 = 1)
    if(('0'&\J\) >= 1) then
        -- set manually
        R(0) <= '1';
        \6R\(2 downto 0) <= "110"; -- 6 x 1
        \18R\(4 downto 0) <= "10010"; -- 18 x 1
        C <= resize(('0'&\J\),C'length) - 1;
    end if;
    -- save 4P-3 for next cycle, 4P = 4, 4-3 = 1
    \4P-3\(0) <= '1';
    -- advance cycle
    i <= i + 1;
    -- position radicand for next cycle
    A <= A(A'high-3 downto 0)&"000";
    state <= COMPUTE_PI;
--
```

```
-- compute next possible P values using R and Pi-1, also precompute ---
-- 8Ci + Ji+1
--
when COMPUTE_PI =>
    -- 4Pi-1 + 18Ri - 3, for when R(0) is 1, limit to MAX_P
    if((\4P-3\ + \18R\) < 0) then
        P_for_R0_1 <= MAX_P;
    else
        P_for_R0_1 <= \4P-3\ + \18R\;
    end if;
    -- 4Pi-1 - 6Ri - 3, for when R(0) is 0, limit to MAX_P
    if((\4P-3\ - \6R\) < 0) then
        P_for_R0_0 <= MAX_P;
    else
        P_for_R0_0 <= \4P-3\ - \6R\;
    end if;
    -- precompute 8Ci and add Ji+1
    \8C&J\ <= (C sll 3) + ('0'&\J\);
    state <= DETERMINE_RI;
--
-- retire the next bit of the result and update the remainder
--
when DETERMINE_RI =>
    -- the last retired bit is required   to determine which
    -- expression is used in creating the new P and R lsb. Partial
    -- products for R used in next cycle are also generated.
    if(R(0) = '1') then
        if(\8C&J\ >= P_for_R0_1) then
            R <= \2R\&'1';
            \6R\ <= x6(\2R\&'1',\6R\'length);
            \18R\ <= x18(\2R\&'1',\18R\'length);
            C <= \8C&J\ - P_for_R0_1;
        else
            R <= \2R\&'0';
            \6R\ <= x6(\2R\&'0',\6R\'length);
            \18R\ <= x18(\2R\&'0',\18R\'length);

            C <= \8C&J\;
        end if;
        -- save 4P for next cycle
        \4P-3\ <= (P_for_R0_1 sll 2) - 3;
    else
        if(\8C&J\ >= P_for_R0_0) then
            R <= \2R\&'1';
```

```vhdl
            \6R\ <= x6(\2R\&'1',\6R\'length);
            \18R\ <= x18(\2R\&'1',\18R\'length);
            C <= \8C&J\ - P_for_R0_0;
        else
            R <= \2R\&'0';
            \6R\ <= x6(\2R\&'0',\6R\'length);
            \18R\ <= x18(\2R\&'0',\18R\'length);
            C <= \8C&J\;
        end if;
        -- save 4P for next cycle
        \4P-3\ <= (P_for_R0_0 sll 2) - 3;
    end if;
    -- position next Ji+1
    A <= A(A'high-3 downto 0)&"000";
    -- check iterations
    if(i < (R'length-1)) then
        i <= i + 1;
        state <= COMPUTE_PI;
    else
        state <= RENORMALIZE;
    end if;
------------------------------------
-- round and renormalize working root
------------------------------------ --
when RENORMALIZE =>
    if(cnt < RENORMALIZE_CNT) then
        R <= '0'&R(R'high downto 1);
        grs(0) <= grs(0) or R(0);
        cnt <= cnt + 1;
    else
        -- include remainder in  sticky
        if(C /= 0) then
            grs(0) <= '1';
        end if;
        -- tranfer bits
        nr <= R(R'high downto 2); -- discard guard and round
        grs(2 downto 1) <= unsigned(R(1 downto 0));
        state <= ROUND;
    end if;
when ROUND =>
    -- round result to nearest even number
    if (rnd = '1' and (grs > 4 or (grs = 4 and nr(0) = '1'))) then
        nr <= nr + 1;
    end if;
```

```vhdl
                    state <= ROUND2;
            -- check for underflow
            when ROUND2 =>
                if(nr = 0) then
                    udr <= '1';
                    busy <= '0';
                    state <= START_EXTR;
                else

                    state <= CONVRT;
                end if;
            -- convert to original sign
            when CONVRT =>
                if(sign = '1') then
                    rt <= not(nr) + 1;
                else
                    rt <= nr;
                end if;
                busy <= '0';
                state <= START_EXTR;
            when others =>
                    state <= RESET;
        end case;
    end if;
end process;

rdy <= '1' when busy = '0' and (extr = '0') else '0';

end RTL;
```

7.2 High Performance Cube Root

Fewer publications are available for cube root than square root, even radix-4 square root. Work published by Naofumi Takagi [3] in 2001 is leveraged here for the algorithm of choice. Takagi leveraged work done by Ercegovac and Lang with some of his own tricks. Because of this continuity there are some similarities with the *Radix-4 Square Root* in the previous chapter section 6.2 and this chapter and section 7.2.

Similarities include the concept of digit-recurrence, the use of a redundant signed-digit set, the use of carry-save adders (keeping the residual in carry-save form), the use of an estimate of the residual and partial result for signed-digit selection, the use of on-the-fly-conversion of the result into standard two's complement, and appendage techniques to eliminate carry propagation.

Dissimilar attributes include radix-2 instead of radix-4, signed-digit set {-1, 0, +1} instead of {-2, -1, 0, +1, +2}, and maintaining a square of the partial result (also in carry-save form).

Because of the added complexity and levels of logic required to implement the algorithm a two-clock pipeline is utilized. Meaning a single bit of the partial result is retired every two clocks.

As with section 6.2 *Radix-4 Square Root*, only specific expressions related to the actual algorithm are referenced, not the supporting arguments or theory. An attempt to use formerly defined terms and variable names is made in the event that the reader wishes to further investigate the premises underlying these expressions.

7.2.1 Architecture Overview

Figures 15 and 16 illustrates the design from a high-level point of view. The first thing to notice is that design has a primary state machine that handles mapping, normalizing, denormalizing, and rounding, while the algorithm processes are handle by three separate but synchronized state machines. Each of the three represents a variable essential to the algorithm's math construct. This partitioning also makes the design more manageable.

C Generation State Machine (C GEN SM) generates the partial result (cube root), *S* Generation State Machine (S GEN SM) generates the square of the partial result (cube root), and *W* Generation State Machine (W GEN SM) computes next residual from the current residual and both *C* and *S* in conjunction with the last retired signed-digit. Then, based on the results of those parameters issues the next signed-digit {-1, 0, +1} for the next iteration.

> Note: *C is maintained in standard two's complement; whereas S and W are maintained in carry-save form.*

The input radicand is first mapped into the internal working register, which are properly scaled to support 3-bit grouping. Then the leading digit is normalized into a fractional value which is less than 1.0 and equal to or greater than 0.001 (1/8). The leading bit is located by left shifting the operand three bits at a time. The number of shifts are recorded for denormalizing the result after the root is extracted.

Next, the normalized value is posted to the *W* Generation State Machine by the primary thread, which sets the initial residual. The initial value for the other state machines are set to the default values required by the algorithm. In essence the initial values represent the first iteration which are always the same except for the initial residual. The primary state machine initiates the algorithm then waits for all three to complete. Subsections that follow provide details of the internal workings of each of the three state machines.

Once all are complete the residual is converted to two's complement form so that the sign can be examined to select either *CP* or *CM*. *CP* is equal to *C* and *CM* is equal to 1 less than *C* (by 1 LSB). This constitutes a correction to the result. It's the same as subtracting 1 from *C* if the residual was negative. The selected value is then denormalized (with recorded *ncnt/fcnt*) by right shifting into the original Qm.n format, then rounded if qualified.

> Note: *cube roots can be negative if the radicand input is negative. The algorithm requires a positive operand and conversion is handled internally. Likewise the resulting root is converted back to negative.*

Below are the basic steps of the algorithm.

Note: *The 3 to 1 ratio typically in a cube root algorithm is not so apparent here due to how the algorithm is structured.*

1). Initialization (also represents iteration 1)

$j = 1$
$C = CP[1] = 2^{-1} = 0.5$
$CM[1] = 0$
$S[1] = 2^{-2} = 0.25$
$W[1] = (2 \bullet \text{initial residual}) - 2^{-2} = (2 \bullet \text{initial residual}) - 0.25$

2). Loop (iterations are the range of bits within the operand)

$\wedge 2 W[j]$ is an estimate of W (residual) truncated to 1 fractional bit, which only requires a 5-bit carry propagation adder. The result qualifies the next signed-digit {-1, 0, +1}.

$$q_{j+1} = \begin{array}{ll} +1 & \text{if } \wedge 2 W[j] \quad > \ 0 \\ 0 & \text{if } \wedge 2 W[j] \quad -0.5 \text{ or } 0 \\ -1 & \text{if } \wedge 2 W[j] \quad \leq \ -1.0 \end{array}$$

$C[j+1]$ = on-the-fly conversion (see C GEN SM)

$S[j+1] = S[j] + 2C[j] \bullet q_{j+1} \bullet 2^{-j-1} + (q_{j+1})^2 \bullet 2^{-2j-2}$

$W[j+1] = 2W[j] -$
$\quad (3S[j] \bullet q_{j+1}) -$
$\quad (3C[j] \bullet (q_{j+1})^2 \bullet 2^{-j-1}) -$
$\quad ((q_{j+1})^3 \bullet 2^{-2j-2})$

The number of iterations is equal to the number of combined integer and fractional bits of the root output Qm.n format.
$j = j + 1$, exit when j = required iterations $- 1$

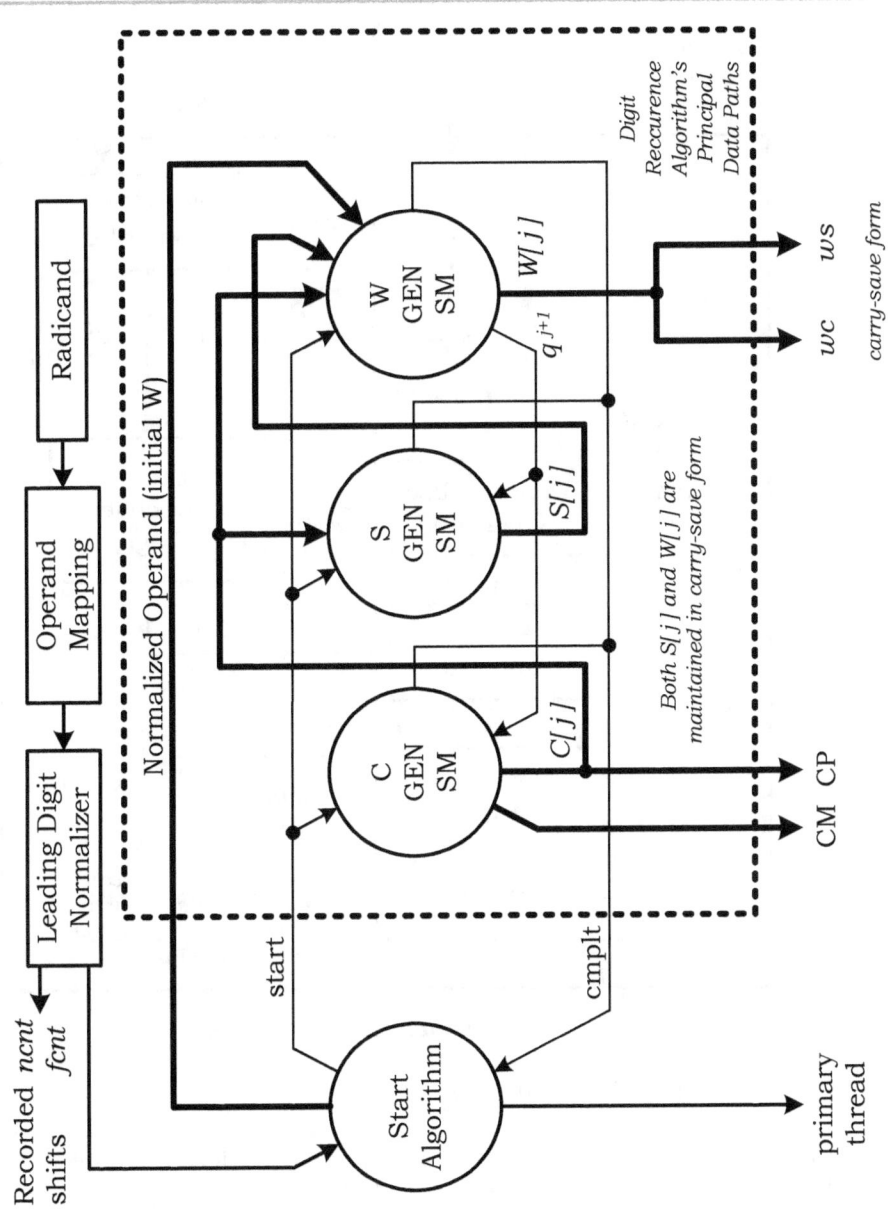

Figure 15

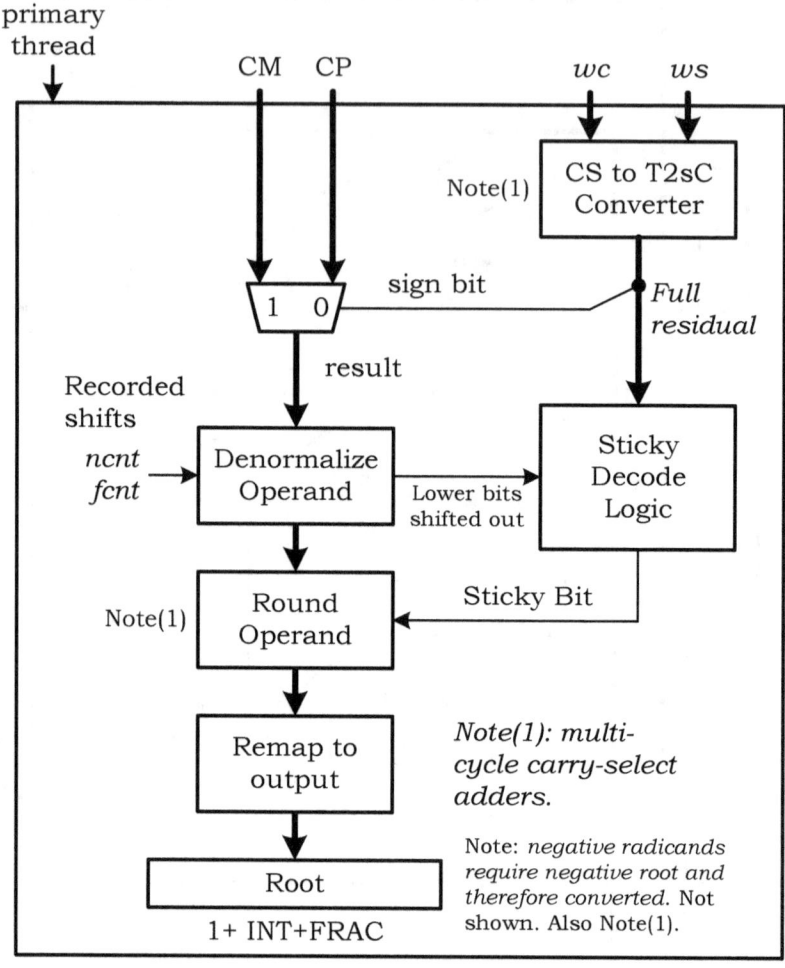

Figure 16

7.2.2 Terms and Variable Names

In the original paper iteration is denoted by the letter j. When within $[j]$ denotes the variables value at that iteration, $[j+1]$ the next value.

q_{j+1} represents the last signed-digit value {-1, 0, +1}
$C[j]$ partial result
$S[j]$ square of the partial result
$W[j]$ residual or remainder

2^{-j-1} or 2^{-2j-2} are single fractional digits used in multiplication

7.2.3 Operand Mapping

Operands used in cube roots need to be in groups of 3-bits apart from the sign bit. This is an imperative for the integer portion but not for the fractional.

The design is scalable for both integer and fractional portions through generics, but internal working registers must be scaled to account for this 3-bit grouping requirement for the integer portion and the bits from the radicand must be properly placed to maintain the original binary-point.

Working registers are temporary registers. Principal registers have the same bit padding but may also have additional integer bits between the sign bit and the implied binary point.

Integer portion

If the integer size is not divisible by 3, one or two extra bits are added to meet this requirement. They are inserted between the MSB and the sign bit. If the integer size is zero or divisible by 3 the same value is used for the internal registers.

Fractional portion

If the fractional size is odd then a bit location is appended to the LSB. Two extra bits are added anyway to serve as guard and round bits, the newly extended LSB will be included in the sticky estimate.

> Note: *the reason the fractional portion can be even is because when the operand is initially left shifted to locate the leading 1-bit zeros can be shift in on the right.*

7.2.4 Leading Digit Normalizer

The MSB of the internal working registers is the sign bit with the implied binary point to its right. The mapped operand is shifted to the left 3-bits at a time until one of the three leading bits is true. This exercise normalizes the operand into a fractional number to satisfy the expression below.

$$1/8 \leq x < 1$$

All seven of theses values qualify 0.001 through 0.111 fall within this range. During shifting two counters are advanced. One counting every triple shift (*ncnt*) and the other counting triple shifts that are within the fractional portion of the operand (*fcnt*). These count values are used to de-normalized the fractional result into the original fixed-point format.

Note: *because dividing by 3 is not a power of 2, a look up table is used during denormalization to divide the recorded count by 3 and properly restore the result to the input/output Qm.n format.*

7.2.5 Principal Registers and Sub Modules

Registers *C*, *S*, and *W* represent the primary variables used in the cube root extraction algorithm. Particulars for each register are provided below followed by a subsection detailing the generation of their corresponding values.

Register C (also denoted as *cr, cpr, cmr*)

- Partial result of the extracted root.
- Formatted as a positive fractional number (1 integer bit for the sign) with an approximate length totaling the number of bits of both integer and fractional portions of the input radicand, plus two bit for rounding.
- In two's complement form.
- A plus and minus, *CP* and *CM*, are maintained for correction purposes prior to rounding. The *CM* is1 LSB less than *CP*. *CP* is equal to *C*.
- Its value is produced by the *C* Generation State Machine.

Register S (also as *ssr* and *scr* for sum and carry components)

- Square of the partial result of the extracted root.
- Formatted as a positive fractional number (1 integer bit for the sign). Its fractional length is twice that of *C*.
- Its value is maintained in carry-save form. Having an equal number of sum and carry bits.
- Its value is produced by the *S* Generation State Machine.

Register W (also as *wsr* and *wcr* for sum and carry components)

- Current residual (remainder) of the last iteration.
- Formatted as a signed fixed-point number having a sign bit plus two integer bits. Its fractional length is twice that of C.
- Its value is maintained in carry-save form. Having an equal number of sum and carry bits.
- Its value is produced by the W Generation State Machine.

7.2.5.1 C Generation State Machine (C GEN SM)

Figure 17 diagrams on-the-fly conversion circuits used to generate
C as well as CP and CM. This technique is commonly used in
divider circuits. Each iteration retires a single bit and the values in
each register may move between each other. The prior CP to CM or
CM to CP depending on the value of signed-digit q_{j+1}.

This technique serves two purposes. First, CM is always 1 LSB less
than the partial result in CP. This is useful when the final residual
(W) of the last iteration is negative. CM will be used as the
extracted root as opposed to CP. Second, the W Generation State
Machine issues a new signed-digit every iteration ranging from {-1,
0, +1}. The idea is that when values can be corrected for on-the-fly.
A digit of {-1} typically means that too much was subtracted from W
and it must be compensated for by subtracting 1 from the resultant
root, hence using CM.

Normally this circuit simply shifts/loads each register, CP and CM,
while shifting in a new value. After all iterations the results are
properly aligned within each corresponding register -- but not so
here.

The value in C is posted to the other state machines to include in
their expressions and therefore must be properly aligned each
iteration as a fractional number. The shift register positions the
incoming bit position for either cp_in or cm_in to be loaded along
with either former CP Reg or CM Reg.

Decode of input bits appended to each register:

$$cp\_in = \begin{cases} 1 \text{ if } & q_{j+1} & +1 \\ 0 \text{ if } & q_{j+1} & 0 \\ 1 \text{ if } & q_{j+1} & -1 \end{cases}$$

$$cm\_in = \begin{cases} 0 \text{ if } & q_{j+1} & +1 \\ 1 \text{ if } & q_{j+1} & 0 \\ 0 \text{ if } & q_{j+1} & -1 \end{cases}$$

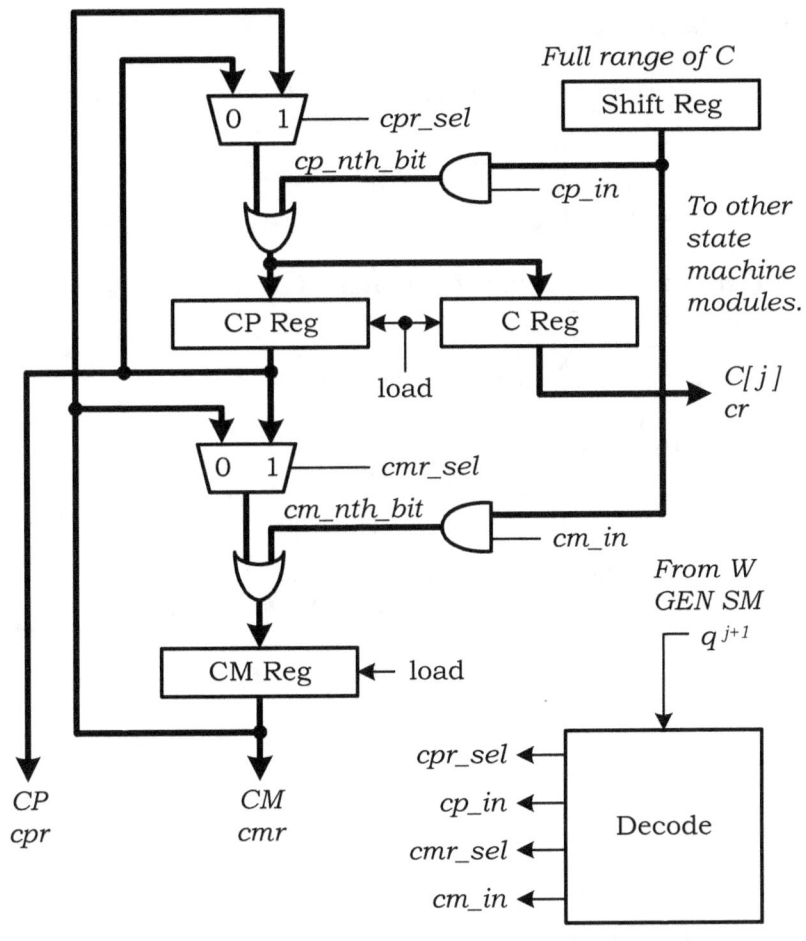

Figure 17

Decode of input mux selection for each register:

$CP[j+1]$ = CP Reg OR'ed with cp_in if $q_{j+1} = 0$ or $+1$
             = CM Reg OR'ed with cp_in if $q_{j+1} = -1$

$CM[j+1]$ = CM Reg OR'ed with cm_in if $q_{j+1} = 0$ or -1
             = CP Reg OR'ed with cm_in if $q_{j+1} = +1$

C GEN SM Sequence

1). Start from primary state machine

$j = 1$
Initialize C, CP, and CM
$cr[1] = cpr[1] = 0.1 = 0.5$
$cmr[1] = 0.0$
Initialize shift-register to first fractional bit location
$sr = 0.10000----0$ binary

2). Wait one clock for next q_{j+1} to be posted

Right shift sr by 1 bit

3). Generate Next C

Decode q_{j+1} and update C/CP and CM registers
$j = j + 1$, exit when iterations reached, otherwise step 2

7.2.5.2 S Generation State Machine (S GEN SM)

The expression for computing the next $S[j]$ value, the square of the partial cube root, is shown below and is maintained in carry-save form.

$$S[j+1] = S[j] + 2C[j] \cdot q_{j+1} \cdot 2^{-j-1} + (q_{j+1})^2 \cdot 2^{-2j-2}$$

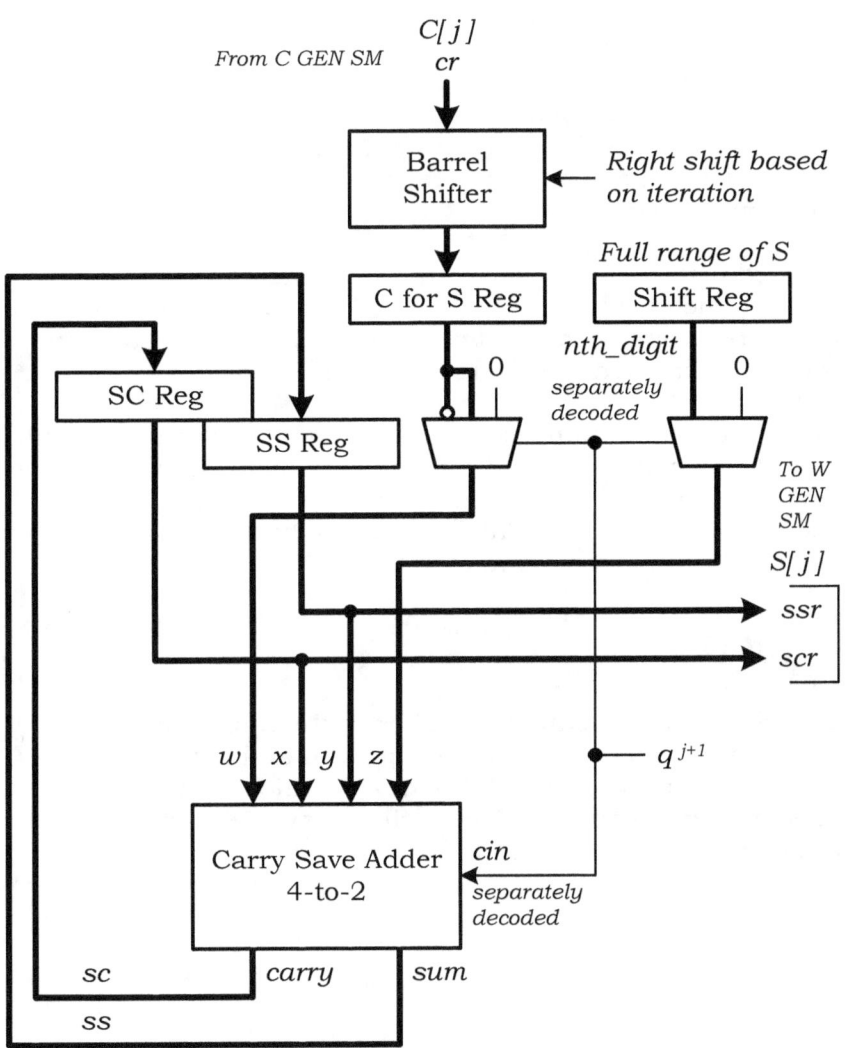

Figure 18

In order to compute $S[j+1]$ the current value of S and C, along with an *nth* digit from a shift register, are input to a 4 to 2 operand carry-save adder, as shown in Figure 18. The S variable is used as is.

The second term multiplies the partial root $C[j]$ by 2, then by q_{j+1} which makes C positive, negative, or zero, and then by a fractional power of 2, 2^{-j-1}.

$$2C[j] \bullet q_{j+1} \bullet 2^{-j-1}$$

One simplification is to combine $2 \bullet 2^{-j-1}$ resulting in half of the fractional power or a fractional power of j only, 2^{-j}. The fractional multiplication of C is the same as right shifting it j number of bits. Each iteration, represented by j, results in the shift being one more bit location to the right.

The signed-digit q_{j+1} determines the resulting polarity of the shifted C, which is registered in C_for_S_Reg. Its output can pass as is, inverted, or zeroed. *Cin* to the carry-save adder is set when the output is inverted, resulting in adding a negative C.

The last term of the expression squares q_{j+1} so its value can only be +1 or 0. Example $+1^2 = 1$, $0^2 = 0$, $-1^2 = 1$. That is multiplied by the fractional portion 2^{-2j-2}, resulting in an *nth_digit* offset to a fractional location or zero. This is the same has shifting an *nth_digit* 2-bits to the right every iteration.

$$(q_{j+1})^2 \bullet 2^{-2j-2}$$

Finally, S, plus the partitioned portion of the expression representing C, plus the partitioned portion of the *nth_digit*, are input to the carry-save adder. The new S, which is in carry-save form is registered and available for the next cycle as well as to the external module W GEN SM.

S GEN SM Sequence

1). Start from primary state machine

$j = 1$
Initialize S
$ssr[1] = 0.01 = 0.25$
$scr[1] = 0$
Initialize shift-register to first fractional bit location
$sr = 0.00010000----0$ binary

2). Generate C for S adder

Barrel right shift $C[j]$ by j and register in c_for_sr

3). Generate Next S

Decode q_{j+1} and update S through carry-save adder, ssr and scr.
Right shift nth_digit by 2-bits

$j = j + 1$, exit when iterations reached, otherwise step 2

7.2.5.3 W Generation State Machine (W GEN SM)

Generation of W is more complex and has terms involving both C and S as well as its own *nth_digit*, as shown in the expression below.

$$W[j+1] = 2W[j] -$$
$$(3S[j] \bullet q_{j+1}) -$$
$$(3C[j] \bullet (q_{j+1})^2 \bullet 2^{-j-1}) -$$
$$((q_{j+1})^3 \bullet 2^{-2j-2})$$

While all functions are contained within a single file, explaining each term requires a different subsection. Figure 15 illustrates the generation of W and the signed-digit q_{j+1} at the highest level, while Figure 19 shows its internals.

The outputs of C GEN SM and S GEN SM modules provide the current $C[j]$ and $S[j]$ values, respectively. Term expressions acting on the these values generate inputs $x1$ through $x4$ into a 6 to 2 operand carry-save adder. Along with expressions acting on the previous $W[j]$ (x5 and x6) a new W is generated $W[j+1]$. Inputs to this adder are maintained in carry-save form for all terms.

After all iterations are complete the residual is passed back to the primary state machine, *wsr* and *wcr* ,where is converted to two's complement to detect whether the residual is negative.

Other particulars include that the primary state machine provides the initial residual value, as well as reads the final residual for converting to two's complement. Next, the q_{j+1} is output to be evaluated to create the next values of C and S. Finally, the *cin_1* and *cin_2* coincide when the S term needs to be subtracted.

W GEN SM Sequence

1). Start from primary state machine
$j = 1$
$q_{j+1} = 0$
Initialize W
$wsr\,[1] = w\_init \times 2$
$wcr\,[1] = 111.11 = -0.25$
Initialize shift-register to first fractional bit location
$sr = 0000.00010000----0$ binary

2). Generate adder inputs

 decode an early q_{j+1} and register as q_{j+1}
 register processed C and S terms
 register processed W term combined with *nth_digit*

3). Generate next W

 register new W discarding MSB
 Right shift *nth_digit* by 2-bits

$j = j + 1$, exit when iterations reached, otherwise step 2

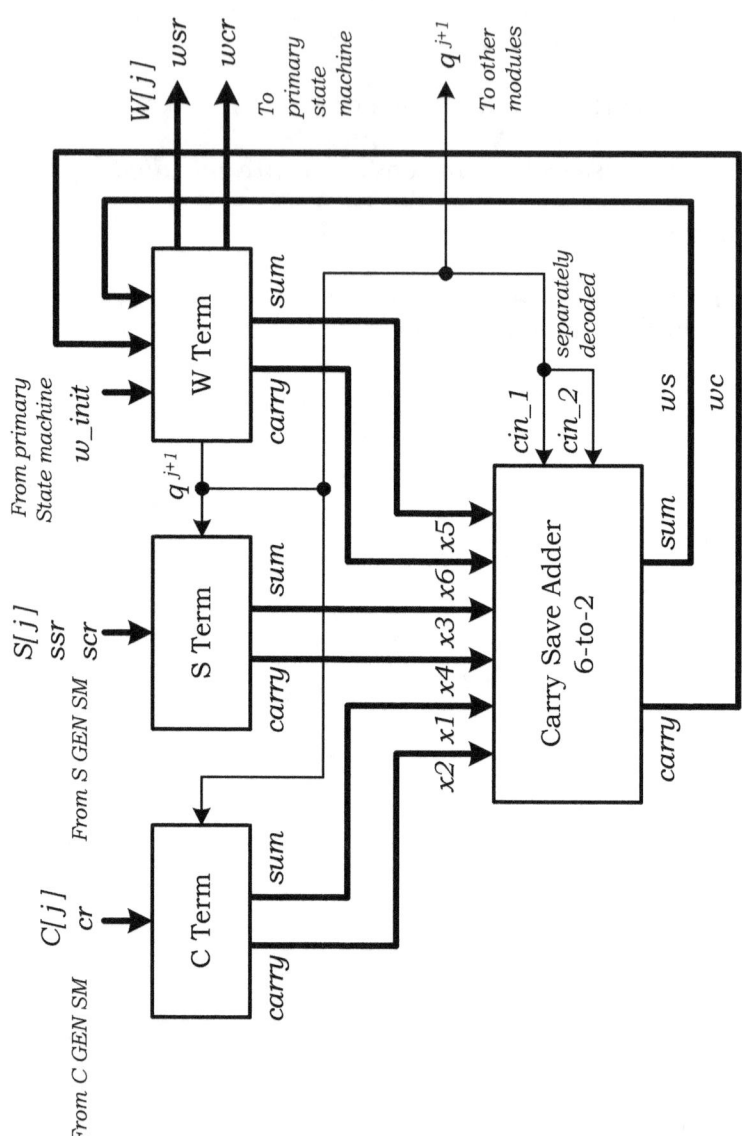

Figure 19

C Term

The term used for processing C has three factors. The third factor multiplies $C[j]$ by a fractional number, 2^{-j-1}, which is the same right shifting its content one more bit each iteration starting with iteration 2, Figure 12.

$$- (3C[j] \bullet (q_{j+1})^2 \bullet 2^{-j-1})$$

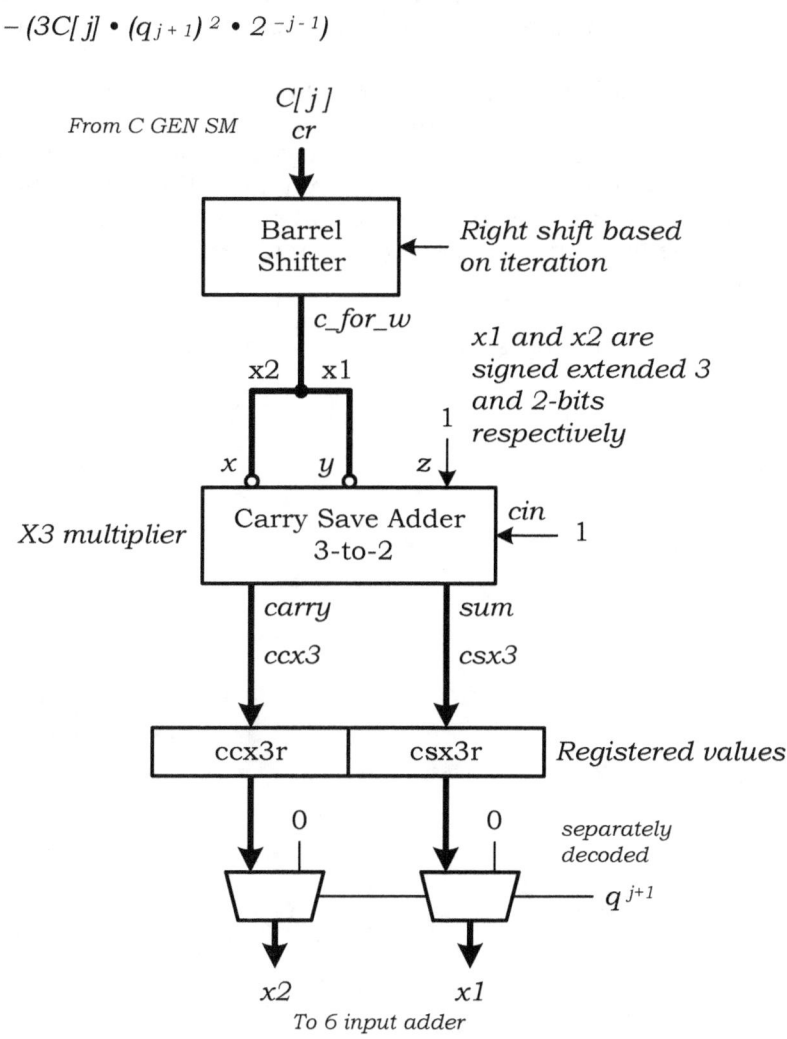

Figure 20

The shifted value, *c_for_w*, is then multiplied by three using a simple partial product technique. Since the multiplier is 2-bits, 0x3, and the multiplicand is a full operand, adding a x1 and x2 (left shifted by 1) accomplishes this. Furthermore each is sign-extended to the full width of the final 6 to 2 carry-save adder.

Note: See chapter 7 in "STATE MACHINES IN VHDL *Multipliers* Vol. 2" for more information on partial product multiplication.

The second factor $(q_{j+1})^2$ means that that the value of *3C[j]* is either multiplied by +1 or 0, never -1. If +1 then the *C* term is subtracted from *W*. Creating a negative *C* term is done by inverting the x1 and x2 components, as well as adding 1 into the *z* input as well as *cin* into the carry-save adder. If q_{j+1} is 0, a zero is selected for the final adder instead of the registered value of the processed *C* term.

Note: *the processed C term is in carry-save form.*

S Term

S is multiplied by 3 by adding both its sum (ssr) and carry (scr) components three times each into a 6 to 2 carry save adder. As shown in Figure 21.

$$- (3S[j] \cdot q_{j+1})$$

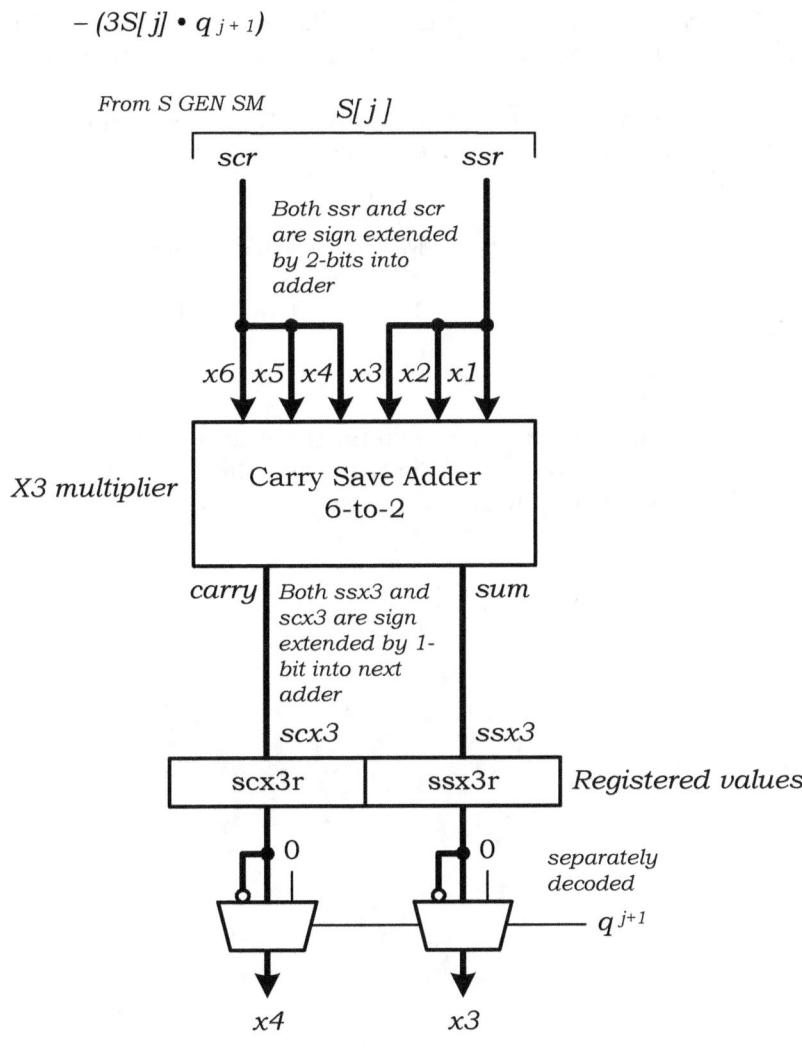

Figure 21

The full expression for W indicates that S is to be subtracted from W, but the S term can be either negative, positive, or zero based on the value of q_{j+1}. If +1 then the processed S term is subtracted from W. If -1 then the S term is added to W. Otherwise a zero is passed and there is no subtraction.

The output multiplexer inverts the S term, passes as is, or passes zero. Not shown here but in Figure 19 are the two *cin* inputs to the final 6 to 2 carry-save adder. When S is to be subtracted *cin_1* and *cin_2* are set to 1. Final step in a two's complement sign conversion.

Since S is in carry-save form, the inputs to the X3 multiplier, as well as its outputs, have to be sign extended. Certain rules have to be followed:

	MSB	MSB	MSB	MSB
sum component	0	1	0	1
carry component	0	0	1	1

All but the third column sign extend the sum while zero padding the carry. If the third column is true then the carry is sign-extended while zero padding the sum.

W Term with nth digit

Processing *W* is combined with its *nth_digit* as shown in Figure 22.

$$2W[j] - ((q_{j+1})^3 \bullet 2^{-2j-2})$$

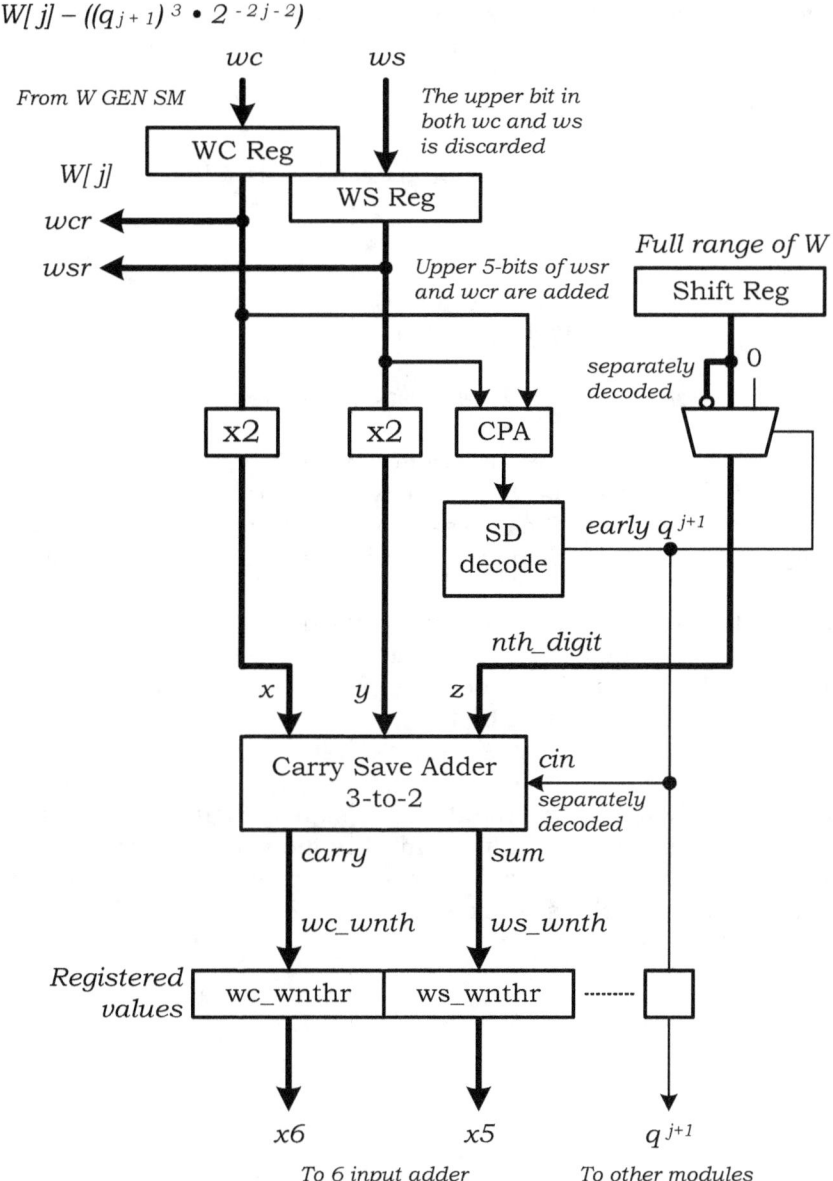

Figure 22

When *W* is updated with a new value from the 6 to 2 carry-save adder, its MSB is discarded (the adder has an extra integer bit). The upper 5-bits of the new *wsr* and *wcr* are used to compute an estimate of *W*. This in turn is used to qualify the next signed-digit. The estimate format as seen by the decode logic is as follows: sign bit, three integer bits and 1 fractional bit.

> w_estimate   Siii.f

The corresponding signed-digit is decoded as follows:

$$q_{j+1} = \begin{cases} +1 & \text{if } \wedge 2\,W[\,j] \quad > \quad 0 \\ 0 & \text{if } \wedge 2\,W[\,j] \quad -0.5 \text{ or } 0 \\ -1 & \text{if } \wedge 2\,W[\,j] \quad \leq \quad -1.0 \end{cases}$$

Note: *evaluating the upper 5-bits is the same as 2W[j].*

The new *wsr* and *wcr* value are left shifted by 1-bit for a x2 with the same width as the final 6 to 2 carry-save adder. However, there is an intermediate 3 to 2 carry-save adder to combine *W* with the next *nth_digit*. Its output being registered as the *W* term into final adder.

The factor $((q_{j+1})^3 \bullet 2^{-2j-2})$ represents the *nth_digit* which is a shift register. Since the cube of q_{j+1} carries the signed-digit's sign forward, the *nth_digit* can be zero, positive, or negative number. Its magnitude is determined by the 2^{-2j-2} factor, which is the same as right shifting the *nth_digit* 2-bits each iteration.

Within the *W* term logic the decoded signed-digit is called early q_{j+1} and only operates within its term. It is, however, registered at the same time as the processed *W* term. It's available for use by the other terms as well as by external modules. In particular C GEN SM and S GEN SM.

7.2.6 Carry Save Form

In conjunction with several variants of carry-save adders carry-save form is utilized throughout this design. See Chapter 9 for details on the different configurations of carry-save adders. Usage in the current design is described below.

- The 3-to-2 carry-save adder has only a single level and is used when computing the C and W terms in the W GEN SM.

- The 4-to-2 variant is used in the S GEN SM.

- The 6-to-2 variant is used in the W GEN SM and when processing the S term in the W GEN SM.

7.2.7 Signed Digit Representation

In most cases signed-digits are encoded with some limited carry information so that when converted to two's complement form their propagation delays are significantly reduced. Signed-digits in this design are used to encode the next sub-operation within the algorithm, namely the update of registers C, S, and W based on the current residual.

The digit-set {-1, 0, +1} are represented by 2-bits each. They are used in conjunction with on-the-fly conversion. Carry propagation is eliminated when updating register C. With S and W registers a digit is used as factors to assign polarity to their respective terms.

Also see section 7.2.5 of "STATE MACHINES IN VHDL Dividers Vol. 3".

7.2.8 Carry Select Adders

See section 7.2.6 of "STATE MACHINES IN VHDL Dividers Vol. 3".

7.2.9 Denormalizing

De-normalization converts the resulting fractional root, which is greater or equal to ½ and less then 1, to the initial fixed-point

Qm.n format. Several parameters determine how many right shifts are required to properly locate the leading data bit relative to the original fixed-point position. Specifically, the internal operand integer and fractional sizes and the recorded shifts during normalization (*ncnt/fcnt*).

The second step is to right shift the internal working register up to 3-bits at a time until the result is properly aligned. As lower bits are shifted they are OR'ed with the sticky bit and not discarded.

Internal Operand Size		*ncnt*	Action taken
integer	fractional		
0	>0	0	No action operand fractional
0	>0	>0	Right shift *ncnt*/3
>0	0	N/A	Right shift Integer size − ((integer size − *ncnt*)/3)
>0	>0	0	Right shift Integer size − (integer size/3)
>0	>0	>0	Normalizing was within integer portion Right shift Integer size − ((integer size − ncnt)/3)
>0	>0	>0	Normalizing was within fractional portion. Right shift Integer size + (*fcnt*/3)

Note: *integer and fractional size are those of the internal operand.*

Note: *General rule -- the root of an integer or a fixed-point number with an integer payload (regardless if there is a fractional content) is smaller than the radicand. In contrast, the root of a fractionally only payload is larger.*

Because cube roots involve 3 to 1 ratios between the radicand and root. Three not being a power of two requires a lookup table to divide the formally saved counts by three.

7.2.10 Remapping

Remapping takes the de-normalized internal register and properly maps each bit into its proper location using the integer and fractional portions of the input/output fixed-point format.

No lower bits are lost in this process but are accounted for in the pre-rounding step.

7.2.11 Correction, Rounding, and Sticky

The result is in two's complement form and is contained in either register *CP or CM* -- register *CM* is *CP* − 1 LSB. The sign of the last residual determines whether the actual result should be *CP* or *CM*. First the carry-save form of the residual, *wsr* and *wcr*, must be added to obtain the signed two's complement equivalent. If the value is positive then register *CP* is the result, if negative *CM* is the result. Furthermore, a positive non-zero residual also sets the sticky bit.

Note: *carry-save to two's complement is converted through a multi-cycle carry-select adder.*

Two extra bits, guard and round, are added to the internal operand registers. These bits along with the LSB of the de-normalized value and sticky bit are used to determine whether the result is rounded. The rounding function can be enabled or disabled at the start of each root extraction cycle.

Note: *rounding is converted through a multi-cycle carry-select adder as well.*

Sticky is set if the residual is a non-zero positive value, or when any lower 1 bits are shifted out during de-normalization, or if the extra LSB bit that was added to make the fractional portion even is set.

Note: *See chapter 5 for Normalizing, Rounding, and Bounds.*

7.2.12 Conclusion

Performance

The number of clock cycles required to complete the operation is approximately the number of bits in the internal working registers minus the sign bit. Additional overhead includes steps for converting negative operands, mapping/remapping, normalizing/de-normalizing (this is data dependent), and rounding.

Note: *See chapter 5 for Normalizing, Rounding, and Bounds.*

Optional Improvements

1). Full operand lengths were employed here to process an ongoing residual to the full length of the input/output fixed-point format.

- Perfect roots however only require a limited number of iterations. The best solution is to add circuits that watch for the residual becoming zero then exit the algorithm early.

  In order to maintain the targeted clock speed, values from *wsr* and *wcr* would have to enter a pipelined adder. This will cause a multi-clock latency but would allow an early result to exit the algorithm sooner.

- Reduce the number of iterations for creating the lower bits. There are no guidelines for this, only that it should not be any less than one third. This is a typical resolution required for cube roots.

- If the input operand data bits only occupy the lower half of the fractional portion, the number of iterations should be INT_SIZ+((FRAC_SIZ/3)). The lower bits would only be shifted out, setting the sticky bit.

2). De-normalization right shifts the fractional result into its former fixed-point position. Increase the number of maximum shifts per cycle to more than 3. If resources are available a barrel shifter could execute in 1 clock.

Example Design

The example state machine provided illustrates the *High Performance Cube Root.*

- Format is scalable fixed-point Qm.n, integer and fractional bit lengths as generic parameters, defaulting to Q31.32, with a range to Q63.64. It is recommend that the combined size of integer and fractional portions be 8 or larger.
- Signed two's complement numbers are supported for the input radicand. All conversions are handled and managed internally by the state machine. A cube root can have a negative result. No imaginary bit is needed.
- Rounding is dynamic, being enabled or disabled during each operation. Defaults to round-to-nearest-even.
- The design has intentional duplication in order to reduce fanout.

As configured for Q31.32, test builds were run using Xilinx ISE.

The following performance was obtained with corresponding part. While a Spartan could be used higher performance parts are recommended.

Xilinx Virtex XC5vlx30 greater than 250MHz

```
--------------------------------------------------------
--
--   HighPrfmCbrt.vhd (High Performance Cube Root)
--
--------------------------------------------------------
library IEEE;
use IEEE.std_logic_1164.all;
use IEEE.numeric_std.all;
-- user packages

entity HighPrfmCbrt is
--
--   Qmn fixed point format is used.
--
--          sign        binary point
--           |              |
--   format <s>(integer bits).(fractional bits)
--            _____/ _____/
--              INT_SIZ        FRAC_SIZ
--
--   Note: full range of operands are supported. The operand length must be
--   odd for integer portion (minus sign) and even for fractional portion. The
--   design scales internal registers to accomplish this.
--
--   However, a minimum of 7-bits should be used, otherwise the design may not
--   operate properly. This is a combined number of integer and fractional
--   bits, which will also set the range of operation.
--
generic (INT_SIZ: integer range 0 to 63 := 31;
         FRAC_SIZ: integer range 0 to 64 := 32);
port
(
     clk: in std_logic; -- system clock
     rst: in std_logic;  -- system reset (must be synchronous)
     -- inputs
     rnd_en: in std_logic; -- enable rounding
     extr: in std_logic; -- initiate root extraction
     rcd: in signed((1+INT_SIZ+FRAC_SIZ)-1 downto 0); -- radicand
     -- outputs
     rdy: out std_logic; -- data ready
     udr: out std_logic; -- underflow
     rt: out signed((1 + INT_SIZ + FRAC_SIZ)-1 downto 0) -- root
);
end HighPrfmCbrt;
```

```
architecture RTL of HighPrfmCbrt is


--------------
-- functions
--------------
function make_groups_of_3_siz(x: integer) return integer is
begin

    if(x = 0) then
        return(0);
    --
    -- if not in groups of 3-bits, adjust
    --
    elsif(x rem 3 = 1) then
        return(x + 2);
    elsif(x rem 3 = 2) then
        return(x + 1);
    --
    -- bits are in groups of 3 already, leave as is
    --
    else
        return(x);
    end if;

end function;

function make_even_siz(x: integer) return integer is
begin

    if(x = 0) then
        return(0);
    -- odd number of bits, so return even
    elsif(x rem 2 /= 0) then
        return(x + 1);
    -- even number of bits, leave as is
    else
        return(x);
    end if;

end function;

function bit_width(val: integer) return integer is
variable bw: integer := 1;
```

```
variable cnt: integer := val;
begin

    while (cnt > 1) loop
        bw := bw+1;
        cnt := cnt/2;
    end loop;
    return(bw);

end function;

-- divide by 3 lookup table initialization
type rom_table_array is array (natural range <>) of natural;
function init_div_by_3_table(entries: natural) return rom_table_array is
variable a: rom_table_array(0 to entries) := (others=>0);
begin
    -- entry 0 already set to 0
    for i in 1 to entries loop
        -- see if divisible by 3
        if(i rem 3 = 0) then
            a(i) := i/3;
        end if;
    end loop;
    return(a);

end function;


---------------------------------
--   type and declared constants
---------------------------------
-- operands integer portion must be divisible by 3 and
-- and the fractional portion even.
constant OP_INT_SIZ: integer := make_groups_of_3_siz(INT_SIZ);
constant OP_FRAC_SIZ: integer := make_even_siz(FRAC_SIZ);

-- sign along with guard and round bits included. if FRAC_SIZ was
-- odd then an extra sticky bit is included in the operand.
constant OP_SIZ: integer := 1 + OP_INT_SIZ + OP_FRAC_SIZ + 2;

-- number of iterations to product final result. The sign is not
-- included, but only the number of bits to fill the output operand
-- plus rounding bits.
constant INTERATIONS: integer := INT_SIZ + FRAC_SIZ + 2;
```

```
-- register scaling

-- size of internal operands of C and associated working
-- registers with sign bit
constant C_REG_SIZ: integer := OP_SIZ;
-- S registers are twice the size of C w/o sign, plus a sign bit
constant S_REG_SIZ: integer := (((OP_SIZ-1)*2)+1);
-- W registers are twice the size of C w/o sign, plus a sign bit and
-- two extra integer bits.
constant W_REG_SIZ: integer := (((OP_SIZ-1)*2)+2+1);


---------------------------------
-- divide by 3 lookup rom table
---------------------------------
constant ROM_TABLE_SIZ: integer := make_groups_of_3_siz(OP_SIZ); --
bit_width(make_groups_of_3_siz(OP_SIZ));
constant DIVIDE_BY_3_TABLE: rom_table_array(0 to ROM_TABLE_SIZ) :=
init_div_by_3_table(ROM_TABLE_SIZ);


---------------
-- components
---------------

component csla is
-- operand size must be greater or equal to partition size
generic (SIZ,PRTN_SIZ: integer);
port
(
    clk: in std_logic;
    rst: in std_logic;
    -- inputs
    wr: in std_logic;
    ci: in std_logic;
    x: in unsigned(SIZ-1 downto 0);
    y: in unsigned(SIZ-1 downto 0);
    -- outputs
    rdy: out std_logic;
    so: out unsigned(SIZ-1 downto 0);
    co: out std_logic
);
end component;

component c_gen_sm is
generic (C_REG_SIZ,INTERATIONS: integer);
```

```vhdl
port
(
    clk: in std_logic;
    rst: in std_logic;
    -- inputs
    start: in std_logic;
    \qj+1\: in signed(1 downto 0);
    -- outputs
    cmplt: out std_logic;
    cr: out unsigned(C_REG_SIZ-1 downto 0);
    cpr: out unsigned(C_REG_SIZ-1 downto 0);
    cmr: out unsigned(C_REG_SIZ-1 downto 0)
);
end component;

component s_gen_sm is
generic (C_REG_SIZ,S_REG_SIZ,INTERATIONS: integer);
port
(
    clk: in std_logic;
    rst: in std_logic;
    -- inputs
    start: in std_logic;
    \qj+1\: in signed(1 downto 0);
    cr: in unsigned(C_REG_SIZ-1 downto 0);
    -- squared partial result, 2C[j]
    cmplt: out std_logic;
    ssr: out unsigned(S_REG_SIZ-1 downto 0);
    scr: out unsigned(S_REG_SIZ-1 downto 0)
);
end component;

component w_gen_sm is
generic (C_REG_SIZ,S_REG_SIZ,W_REG_SIZ,INTERATIONS: integer);
port
(
    clk: in std_logic; -- system clock
    rst: in std_logic;  -- system reset
    -- inputs
    start: in std_logic;
    w_init: in unsigned(W_REG_SIZ-1 downto 0);
    cr: in unsigned(C_REG_SIZ-1 downto 0);
    scr: in unsigned(S_REG_SIZ-1 downto 0);
    ssr: in unsigned(S_REG_SIZ-1 downto 0);
```

```vhdl
    -- current residual
    cmplt: out std_logic;
    wsr: out unsigned(W_REG_SIZ-1 downto 0);
    wcr: out unsigned(W_REG_SIZ-1 downto 0);
    \c qj+1\: out signed(1 downto 0);
    \s qj+1\: out signed(1 downto 0)
);
end component;


----------------------
--   declared signals
----------------------
-- internal sign conversions registers with input/output supporting
-- radicand and root sizes.
signal tmp1: unsigned(rcd'length-1 downto 0) := (others=>'0');
-- conversion adder
signal adder1_in1: unsigned(rcd'length-1 downto 0) := (others=>'0');
signal adder1_in2: unsigned(rcd'length-1 downto 0) := (others=>'0');
signal adder1_out: unsigned(rcd'length-1 downto 0) := (others=>'0');
signal rcd_sign: std_logic := '0';
signal adder1_wr: std_logic := '0';
signal adder1_rdy: std_logic := '0';


-- size of internal operands of C and associated working registers
signal nr: unsigned(C_REG_SIZ-1 downto 0) := (others=>'0');
signal cr: unsigned(C_REG_SIZ-1 downto 0) := (others=>'0');
signal cpr: unsigned(C_REG_SIZ-1 downto 0) := (others=>'0');
signal cmr: unsigned(C_REG_SIZ-1 downto 0) := (others=>'0');
signal c_cmplt: std_logic := '0';
signal tmp2: unsigned(C_REG_SIZ-1 downto 0) := (others=>'0');
signal tmp3: unsigned(C_REG_SIZ-1 downto 0) := (others=>'0');
signal tmp4: unsigned(C_REG_SIZ-1 downto 0) := (others=>'0');


-- internal register rounding adder
signal adder2_in1: unsigned(C_REG_SIZ-1 downto 0) := (others=>'0');
signal adder2_in2: unsigned(C_REG_SIZ-1 downto 0) := (others=>'0');
signal adder2_out: unsigned(C_REG_SIZ-1 downto 0) := (others=>'0');
signal adder2_wr: std_logic := '0';
signal adder2_rdy: std_logic := '0';


-- S registers
signal ssr: unsigned(S_REG_SIZ-1 downto 0) := (others=>'0');
signal scr: unsigned(S_REG_SIZ-1  downto 0) := (others=>'0');
signal s_cmplt: std_logic := '0';
```

```
-- W registers
signal w_init: unsigned(W_REG_SIZ-1 downto 0) := (others=>'0');
signal wsr: unsigned(W_REG_SIZ-1 downto 0) := (others=>'0');
signal wcr: unsigned(W_REG_SIZ-1 downto 0) := (others=>'0');
signal w_cmplt: std_logic := '0';

-- residual (W) two's complement conversion adder
signal adder3_in1: unsigned(wsr'length-1 downto 0) := (others=>'0');
signal adder3_in2: unsigned(wsr'length-1 downto 0) := (others=>'0');
signal adder3_out: unsigned(wsr'length-1 downto 0) := (others=>'0');
signal adder3_wr: std_logic := '0';
signal adder3_rdy: std_logic := '0';

-- the plus 2 is because the fractional portion is even
signal ncnt: integer range 0 to nr'length+2 := 0;
signal fcnt: integer range 0 to nr'length+2 := 0;
signal dcnt: integer range 0 to nr'length+2 := 0;
signal portion: std_logic := '0';
signal grs: unsigned(2 downto 0) := "000";
signal lsb: std_logic := '0';

-- rom table address and data
signal to_divide_by_3: integer range 0 to nr'length+2 := 0;
signal divided_by_3 : integer range 0 to nr'length+2 := 0;

signal start: std_logic := '0';
signal \c qj+1\: signed(1 downto 0) := (others=>'0');
signal \s qj+1\: signed(1 downto 0) := (others=>'0');

-----------------------
--   enumeration lists
-----------------------
type sm_def is
(
    RESET,
    START_CBRT,
    CNVRT_TO_POS,
    CNVRT_TO_POS2,
    MAP_BITS,
    NORMALIZE_OP,
    START_ALGO,
    REM_COR,
    REM_COR2A,
```

```
    REM_COR2B,
    REM_COR2C,
    DENORMALIZE,
    DENORMALIZE2A,
    DENORMALIZE2B,
    DENORMALIZE2C,
    DENORMALIZE3,
    PREROUND,
    ROUND,
    ROUND2,
    COMPLETE,
    COMPLETE2,
    COMPLETE3,
    COMPLETE4,
    ERROR,
    WAIT_ONE_CLK
);
signal state, ret_state: sm_def := RESET;

--------------------------------- module code ----------------------------
begin


-----------------------------------------------
--
--   HighPrfmCbrt Cube Root master state machine
--
-----------------------------------------------
process(rst,clk)
begin

    if(rst='1') then

        -- working registers
        tmp1 <= (others=>'0');
        adder1_in1 <= (others=>'0');
        adder1_in2 <= (others=>'0');
        adder2_in1 <= (others=>'0');
        adder2_in2 <= (others=>'0');
        adder3_in1 <= (others=>'0');
        adder3_in2 <= (others=>'0');
        w_init <= (others=>'0');
        nr <= (others=>'0');
        tmp1 <= (others=>'0');
```

STATE MACHINES IN VHDL *Root Functions* Vol. 5

```vhdl
        tmp2 <= (others=>'0');
        tmp3 <= (others=>'0');
        tmp4 <= (others=>'0');

        -- misc
        rcd_sign <= '0';
        udr <= '0';
        portion <= '0';
        ncnt <= 0;
        fcnt <= 0;
        grs <= (others=>'0');
        lsb <= '0';
        rt <= (others=>'0');
        to_divide_by_3 <= 0;

        -- control
        adder1_wr <= '0';
        adder2_wr <= '0';
        adder3_wr <= '0';
        start <= '0';

        -- states
        state <= RESET;
        ret_state <= RESET;

elsif rising_edge(clk) then

        -- one clock signals
        adder1_wr <= '0';
        adder2_wr <= '0';
        adder3_wr <= '0';


        --
        --   state machine body
        --
        case state is
            -- reset state
            when RESET =>
                state <= START_CBRT;
                --
                --   cube root body
                --
            when START_CBRT =>
                -- wait for extraction command
```

```
    if(extr = '1') then
        udr <= '0';
        tmp1 <= unsigned(rcd);
        ncnt <= 0;
        fcnt <= 0;
        portion <= '0';
        state <= CNVRT_TO_POS;
    end if;
--
-- convert radicand to positive operand using fast adder.
--
when CNVRT_TO_POS =>
    if(tmp1(tmp1'high)  = '1') then
        rcd_sign <= '1'; -- keep tract of sign
        adder1_in1 <= not(tmp1);
        adder1_in2 <= to_unsigned(1,adder1_in2'length);
        adder1_wr <= '1';
        state <= CNVRT_TO_POS2;
    else
        state <= MAP_BITS;
    end if;
when CNVRT_TO_POS2 =>
    if(adder1_rdy = '1') then
        tmp1 <= adder1_out;
        state <= MAP_BITS;
    end if;
--
-- map radicand into internal operand that will act as initial residual W[0]
--
when MAP_BITS =>
    -- map fractional bits
    if(FRAC_SIZ /= 0) then
        -- when equal just include guard and round
        if(OP_FRAC_SIZ = FRAC_SIZ) then
            nr((FRAC_SIZ+2)-1 downto 0) <=
                tmp1(FRAC_SIZ-1 downto 0)&"00";
        -- or larger by 1 to account for extra bit
        else
            nr((FRAC_SIZ+3)-1 downto 0) <=
                tmp1(FRAC_SIZ-1 downto 0)&"000";
        end if;
    end if;
    -- map integer bits
    if(INT_SIZ /= 0) then
```

```vhdl
                    -- when equal (3-bit groups must be maintained)
                    if(OP_INT_SIZ = INT_SIZ) then
                        nr(nr'high downto OP_FRAC_SIZ+2) <=
                            tmp1(tmp1'high downto FRAC_SIZ);
                    -- larger by 1 or 2 bits sign extend
                    else
                        -- larger by 1 bit
                        if(OP_INT_SIZ - INT_SIZ = 1) then
                            nr(nr'high downto OP_FRAC_SIZ+2) <=
                                '0'&tmp1(tmp1'high downto FRAC_SIZ);
                        -- larger by 2 bits
                        else
                            nr(nr'high downto OP_FRAC_SIZ+2) <=
                                "00"&tmp1(tmp1'high downto FRAC_SIZ);
                        end if;
                    end if;
                else
                    nr(nr'high) <= '0';  -- sign bit is always positive

            end if;
            state <= NORMALIZE_OP;
    --
    -- normalize operand to between 1/8 and 7/8 (< 1) by left
    -- shifting 3-bits at a time. if the entire content is zero its an
    -- error. ignore sign bit, it should always be zero.
    --
    -- every counts represents a 3-bit shift.
    --
    when NORMALIZE_OP =>
        if(nr(nr'high-1 downto nr'high - 3) /= "000") then
                -- initial W value observing relative binary point
                w_init <= (others=>'0'); -- clear all lower bits
                w_init(w_init'high downto w_init'high-nr'length-1) <=
                    "00"&nr(nr'high downto 0);
                start <= '1';
                state <= START_ALGO;
        -- check for limit of operand
        elsif(ncnt < nr'length) then
            ncnt <= ncnt + 3;
            nr(nr'high-1 downto 0) <= nr(nr'high-4 downto 0)&"000";

        -- lower boundary reached, operand is zero
        else
            state <= ERROR;
```

```
        end if;
        -- determine portion of operand where leading digits are
        if(ncnt = OP_INT_SIZ) then
            portion <= '1';
        elsif(ncnt > OP_INT_SIZ) then
            fcnt <= fcnt + 3;
        end if;
    --
    -- algorithm started, wait for other state machines to complete
    --
    when START_ALGO =>
        if(c_cmplt = '1' and s_cmplt = '1' and w_cmplt = '1') then
            -- capture registers
            adder3_in1 <= wsr; -- residual
            adder3_in2 <= wcr;
            tmp2 <= cpr; -- result positive and positive -1
            tmp3 <= cmr;
            -- clear handshake
            start <= '0';
            adder3_wr <= '1';
            state <= REM_COR;
        end if;
    --
    -- Remainder and root correction/conversion
    --
    when REM_COR =>
        -- wait for adder to complete
        if(adder3_rdy = '1') then
            -- check to see if residual is negative
            if(adder3_out(adder3_out'high) = '1') then
                state <= REM_COR2C;
            -- residual zero
            elsif(adder3_out = 0) then
                state <= REM_COR2A;
            -- residual positive
            else
                state <= REM_COR2B;
            end if;
        end if;
    -- the residual was zero
    when REM_COR2A =>
        tmp4 <= unsigned(tmp2);--cpr
        grs(0) <= '0';
        state <= DENORMALIZE;
```

```
-- the residual was positive
when REM_COR2B =>
    tmp4 <= unsigned(tmp2);
    grs(0) <= '1';
    state <= DENORMALIZE;
-- the residual was negative (too far)
when REM_COR2C =>
    tmp4 <= unsigned(tmp3);
    grs(0) <= '0';
    state <= DENORMALIZE;
--
--   Final steps
--
when DENORMALIZE =>
    -- fractionally only
    if(OP_INT_SIZ = 0 and OP_FRAC_SIZ > 0) then
        -- already aligned
        if(ncnt = 0) then
            dcnt <= 0;
            state <= DENORMALIZE3;
        -- fractionally unaligned, ncnt/3
        else
            to_divide_by_3 <= ncnt;
            state <= WAIT_ONE_CLK;
            ret_state <= DENORMALIZE2A;
        end if;
    -- integer only
    elsif(OP_INT_SIZ > 0 and OP_FRAC_SIZ = 0) then
        to_divide_by_3 <= (OP_INT_SIZ - ncnt);
        state <= WAIT_ONE_CLK;
        ret_state <=   DENORMALIZE2B;
    -- integer and fractional portions
    else
        -- upper integer bits set
        if(ncnt = 0) then
            dcnt <= OP_INT_SIZ - OP_INT_SIZ/3;
            state <= DENORMALIZE3;
        -- full range of values
        else
            -- within integer portion
            if(portion = '0') then
                to_divide_by_3 <= (OP_INT_SIZ - ncnt);
                state <= WAIT_ONE_CLK;
                ret_state <=   DENORMALIZE2B;
```

```vhdl
                    -- within fractional portion
                    else
                        to_divide_by_3 <= fcnt;
                        state <= WAIT_ONE_CLK;
                        ret_state <= DENORMALIZE2C;
                    end if;
                end if;
            end if;
    -- dcnt = ncnt/3
    when DENORMALIZE2A =>
        dcnt <= divided_by_3;
        state <= DENORMALIZE3;
    -- dcnt = op_int_siz - (op_int_siz - ncnt)/3
    when DENORMALIZE2B =>
        dcnt <= OP_INT_SIZ - divided_by_3;
        state <= DENORMALIZE3;
    -- dcnt = op_int_siz + fcnt/3
    when DENORMALIZE2C =>
        dcnt <= OP_INT_SIZ + divided_by_3;
        state <= DENORMALIZE3;

    -- do it fast
    when DENORMALIZE3 =>
        if(dcnt = 0) then
            state <= PREROUND;
        elsif(dcnt = 1) then
            dcnt <= dcnt - 1;
          tmp4(tmp4'high) <= '0';
            tmp4(tmp4'high-1 downto 0) <= "0"&tmp4(tmp4'high-1 downto 1);
            -- set sticky if any lower true bits are discarded
            grs(0) <= grs(0) or tmp4(0);
            state <= PREROUND;
        elsif(dcnt = 2) then
            dcnt <= dcnt - 2;
            tmp4(tmp4'high) <= '0';
            tmp4(tmp4'high-1 downto 0) <=
                "00"&tmp4(tmp4'high-1 downto 2);
            -- set sticky if any lower true bits are discarded
            grs(0) <= grs(0) or tmp4(1) or tmp4(0);

            state <= PREROUND;
        else -- dcnt > 2
            dcnt <= dcnt - 3;
            tmp4(tmp4'high) <= '0';
```

```
        tmp4(tmp4'high-1 downto 0) <=
            "000"&tmp4(tmp4'high-1 downto 3);
        -- set sticky if any lower true bits are discarded
        grs(0) <= grs(0) or tmp4(2) or tmp4(1) or tmp4(0);
    end if;
-- prepare operand for rounding, extract guard and round
when PREROUND =>
    if(FRAC_SIZ /= 0) then
        -- no extra bit
        if(OP_FRAC_SIZ = FRAC_SIZ) then
            tmp2 <= tmp4(tmp4'high downto 2)&"00";
            grs(2 downto 1) <= tmp4(1 downto 0);
            lsb <= tmp4(2);
        -- or extra bit with sticky
        else
            tmp2 <= tmp4(tmp4'high downto 3)&"000";
            grs(2 downto 1) <= tmp4(2 downto 1);
            grs(0) <= grs(0) or tmp4(0);
            lsb <= tmp4(3);
        end if;
    -- no fractional portion
    else
            tmp2 <= tmp4(tmp4'high downto 2)&"00";
            grs(2 downto 1) <= tmp4(1 downto 0);
            lsb <= tmp4(2);
    end if;
    state <= ROUND;
-- check if rounding is enabled, then do so based on rules
when ROUND =>
    if(rnd_en = '1' and(grs > 4 or (grs = 4 and lsb = '1'))) then
        adder2_in1 <= tmp2;
        -- account for guard and round offset
        if(FRAC_SIZ /= 0) then
                -- offset different with extra bit
                if(OP_FRAC_SIZ = FRAC_SIZ) then

                    adder2_in2 <= to_unsigned(4,adder2_in2'length);
                else
                    adder2_in2 <= to_unsigned(8,adder2_in2'length);
                end if;
        -- no fractional portion
        else
            adder2_in2 <= to_unsigned(4,adder2_in2'length);
        end if;
```

```vhdl
                adder2_wr <= '1';
                state <= ROUND2;
            else
                tmp3 <= tmp2;
                state <= COMPLETE;
            end if;
    when ROUND2 =>
        if(adder2_rdy = '1') then
            tmp3 <= adder2_out;
            state <= COMPLETE;
        end if;
    --
    -- remap bit and convert root to negative if radicand was negative
    --
    when COMPLETE =>
        -- map fractional bits
        if(FRAC_SIZ /= 0) then
            -- transfer over, minus rounding bits
            if(OP_FRAC_SIZ = FRAC_SIZ) then
                tmp1(FRAC_SIZ-1 downto 0) <=
                    tmp3((FRAC_SIZ+2)-1 downto 2);
            -- or larger by 1 account for extra bit
            else
                tmp1(FRAC_SIZ-1 downto 0) <=
                    tmp3((FRAC_SIZ+3)-1 downto 3);
            end if;
        end if;
        -- map sign and integer bits
        if(INT_SIZ /= 0) then
            -- account for two rounding bits
            if(OP_FRAC_SIZ = FRAC_SIZ) then
                tmp1(INT_SIZ+FRAC_SIZ-1 downto FRAC_SIZ) <=
                    tmp3((INT_SIZ+FRAC_SIZ+2)-1 downto FRAC_SIZ+2);
            -- account for extra bit
            else
                tmp1(INT_SIZ+FRAC_SIZ-1 downto FRAC_SIZ) <=
                    tmp3((INT_SIZ+FRAC_SIZ+3)-1 downto FRAC_SIZ+3);
            end if;
        end if;
        -- sign is always positive
        tmp1(rt'high) <= '0';
        state <= COMPLETE2;
    -- check if radicand was negative
    when COMPLETE2 =>
```

```vhdl
                if(rcd_sign = '1') then
                    adder1_in1 <= not(tmp1);
                    adder1_in2 <= to_unsigned(1,adder1_in2'length);
                    adder1_wr <= '1';
                    state <= COMPLETE3;
                else
                    state <= COMPLETE4;
                end if;
            when COMPLETE3 =>
                if(adder1_rdy = '1') then
                    rt <= signed(adder1_out);
                    state <= START_CBRT;
                end if;
            when COMPLETE4 =>
                rt <= signed(tmp1);
                state <= START_CBRT;
            --
            --  error occurred
            --
            when ERROR =>
                udr <= '1';
                state <= START_CBRT;
            --
            --  wait for 1 clock
            --
            when WAIT_ONE_CLK =>
                state <= ret_state;

            when others => state <= RESET;
        end case;
    end if;

end process;

rdy <= '1' when (extr = '0') and (state = START_CBRT) else '0';


----------------------------------------
--
-- synchronous ROM (inferred READ-FIRST)
--
----------------------------------------
process(clk)
begin
```

```
    if rising_edge(clk) then
        divided_by_3 <= DIVIDE_BY_3_TABLE(to_divide_by_3 );
    end if;

end process;


-----------------------------------------------
--
--   adder 1 for two's complement sign conversion
--
-----------------------------------------------
--   General use Carry Select Adder (fast adder)
csla_mod1: csla
generic map (SIZ => rcd'length, PRTN_SIZ => 8)
port map
(
    clk => clk,
    rst => rst,
    -- inputs
    wr => adder1_wr,
    ci => '0',
    x => adder1_in1,
    y => adder1_in2,
    -- outputs
    rdy => adder1_rdy,
    so => adder1_out,
    co => open
);


----------------------------------------------------------
--
--   adder 2 for rounding result
--
----------------------------------------------------------
--   General use Carry Select Adder (fast adder)
csla_mod2: csla
generic map (SIZ => OP_SIZ, PRTN_SIZ => 8)
port map
(
    clk => clk,
    rst => rst,
    -- inputs
    wr => adder2_wr,
    ci => '0',
```

```
    x => adder2_in1,
    y => adder2_in2,
    -- outputs
    rdy => adder2_rdy,
    so => adder2_out,
    co => open
);


-------------------------------------------------------
--
--   adder 2 for converting residual into two's complement
--
-------------------------------------------------------
--   General use Carry Select Adder (fast adder)
csla_mod3: csla
generic map (SIZ => wsr'length, PRTN_SIZ => 8)
port map
(
    clk => clk,
    rst => rst,
    -- inputs
    wr => adder3_wr,
    ci => '0',
    x => adder3_in1,
    y => adder3_in2,
    -- outputs
    rdy => adder3_rdy,
    so => adder3_out,
    co => open
);



----------------------------------------------------------------
--
--   generation of C[j], denoted by cr, which is the partial result
--
----------------------------------------------------------------
c_gen_sm_mod: c_gen_sm
generic map(C_REG_SIZ => C_REG_SIZ, INTERATIONS => INTERATIONS)
port map
(
    clk => clk,
    rst => rst,
    -- inputs
```

```
        start => start,
        \qj+1\ => \c qj+1\,
        -- outputs
        cmplt => c_cmplt,
        cr => cr,
        cpr => cpr,
        cmr => cmr
);


    ------------------------------------------------------------------
    --
    --   generation of S[j], denoted by ssr and scr in carry-save form,is
    --   the square of the partial result C[j]
    --
    ------------------------------------------------------------------
    s_gen_sm_mod: s_gen_sm
    generic map(C_REG_SIZ => C_REG_SIZ, S_REG_SIZ => S_REG_SIZ,
                        INTERATIONS =>  INTERATIONS)
    port map
    (
        clk => clk,
        rst => rst,
        -- inputs
        start => start,
        \qj+1\ => \s qj+1\,
        cr => cr,
        -- outputs
        cmplt => s_cmplt,
        ssr => ssr,
        scr => scr
);


    ------------------------------------------------------------------
    --
    --   generation of W[j], denoted by wsr and wcr in carry-save form, is
    --   the residual (remainder).
    --
    ------------------------------------------------------------------
    w_gen_sm_mod: w_gen_sm
    generic map(C_REG_SIZ => C_REG_SIZ, S_REG_SIZ => S_REG_SIZ,
                        W_REG_SIZ => W_REG_SIZ, INTERATIONS =>
    INTERATIONS)
    port map
    (
```

```
        clk => clk,
        rst => rst,
        -- inputs
        start => start,
        w_init => w_init,
        cr => cr,
        scr => scr,
        ssr => ssr,
        -- current residual
        cmplt => w_cmplt,
        wsr => wsr,
        wcr => wcr,
        \c qj+1\ => \c qj+1\ ,
        \s qj+1\ => \s qj+1\
    );

end RTL;
```

```
------------------------------------------------------------
--
--   csla.vhd Carry Select Adder module
--
-- Multi-stage carry select adder. Number of stages is based
-- on the operand size and the number of partitions needed
-- to support the operand size.
------------------------------------------------------------
library IEEE;
use IEEE.std_logic_1164.all;
use IEEE.numeric_std.all;

entity csla is
-- operand size must be greater or equal to partition size
generic (SIZ,PRTN_SIZ: integer);
port
(
    clk: in std_logic; -- system clock
    rst: in std_logic;  -- system reset (must be synchronous)
    -- inputs
    wr: in std_logic;
    ci: in std_logic;
    x: in unsigned(SIZ-1 downto 0);
    y: in unsigned(SIZ-1 downto 0);
    -- outputs
    rdy: out std_logic;
    so: out unsigned(SIZ-1 downto 0);
    co: out std_logic
);
end csla;

architecture RTL of csla is

-- functions
function gen_total_seg_reg(extra,base: integer) return integer is
variable cnt: integer := base;
begin
    if(extra > 0) then
        cnt := cnt + 1;
    end if;
    return(cnt);
end function;

-- data segmenting
```

```vhdl
type seg_type is array (integer range <>) of unsigned(PRTN_SIZ-1 downto 0);

-- sizing constants
constant EXTRA_BITS: integer := SIZ rem PRTN_SIZ;
constant NUM_SEG_REGS: integer := SIZ / PRTN_SIZ;
constant TOTAL_SEG_REG: integer :=
gen_total_seg_reg(EXTRA_BITS,NUM_SEG_REGS);

-- local signals
signal a,b,s: unsigned(x'length-1 downto 0) := (others=>'0');
signal c,c_reg: unsigned(TOTAL_SEG_REG-1 downto 0) :=  (others=>'0');
signal rdy_dly: unsigned(TOTAL_SEG_REG-1 downto 0) := (others=>'0');

-- test signals
signal s0,s1: seg_type(1 to NUM_SEG_REGS-1) := (others=>(others=>'0'));
signal c0,c1: unsigned(1 to NUM_SEG_REGS-1) := (others=>'0');

----------------------------------- module code -----------------------------
begin

-- register process
process(rst,clk)
begin

    if(rst='1') then
        a <= (others=>'0');
        b <= (others=>'0');
        c_reg <= (others=>'0');
        so <= (others=>'0');
        rdy_dly <= (others=>'1');
    elsif rising_edge(clk) then
        -- input operands
        if(wr = '1') then
            a <= x;
            b <= y;
        end if;
        -- register selected sums and carries
        c_reg <= c;
        so <= s;
        -- ready delay logic
        if(wr = '1') then
            rdy_dly <= (others=>'0');
        else
            rdy_dly <= rdy_dly(rdy_dly'high-1 downto 0)&'1';
```

```
        end if;
      end if;

end process;
-- last carry in chain is output
co <= c_reg(c_reg'high);
rdy <= rdy_dly(rdy_dly'high) when wr = '0' else '0';

-- base segment adder
base_seg:entity work.cpa generic map(PRTN_SIZ)
port map
(   ci, -- carry in from parent module
    a(PRTN_SIZ-1 downto 0),
    b(PRTN_SIZ-1 downto 0),
    s(PRTN_SIZ-1 downto 0),
    c(0)
);

-- adders for upper segments (minus base segment)
gen_upr_seg: if NUM_SEG_REGS > 1 generate
--signal s0,s1: seg_type(1 to NUM_SEG_REGS-1) := (others=>(others=>'0'));
--signal c0,c1: unsigned(1 to NUM_SEG_REGS-1) := (others=>'0');
begin

    -- create adders and interconnecting muxes
    gen_upr_adr: for i in 1 to NUM_SEG_REGS-1 generate
        -- compute segment for each carry input
        upr_cpa_0:entity work.cpa generic map(PRTN_SIZ)
        port map
        (   '0',-- carry in is = 0
            a((PRTN_SIZ*i)+PRTN_SIZ-1 downto (PRTN_SIZ*i)),
            b((PRTN_SIZ*i)+PRTN_SIZ-1 downto (PRTN_SIZ*i)),
            s0(i),c0(i)
        );
        upr_cpa_1:entity work.cpa generic map(PRTN_SIZ)
        port map
        (   '1',-- carry in is = 1
            a((PRTN_SIZ*i)+PRTN_SIZ-1 downto (PRTN_SIZ*i)),
            b((PRTN_SIZ*i)+PRTN_SIZ-1 downto (PRTN_SIZ*i)),
            s1(i),c1(i)
        );
        -- select sum and carries for current segment, based on previous carry
        c(i) <= c0(i) when c_reg(i-1) = '0' else c1(i);
        s((PRTN_SIZ*i)+PRTN_SIZ-1 downto (PRTN_SIZ*i)) <=
```

```
        s0(i) when c_reg(i-1) = '0' else s1(i);
    end generate;

end generate;

-- extra-bits adder
gen_xtr_bits: if EXTRA_BITS > 0 generate
signal s0,s1: unsigned(EXTRA_BITS-1 downto 0) := (others=>'0');
signal c0,c1: std_logic := '0';
begin

    xtr_cpa_0:entity work.cpa generic map(EXTRA_BITS)
    port map
    (   '0',-- carry in is = 0
        a(a'high downto (PRTN_SIZ*NUM_SEG_REGS)),
        b(b'high downto (PRTN_SIZ*NUM_SEG_REGS)),
        s0,c0
    );
    xtr_cpa_1:entity work.cpa generic map(EXTRA_BITS)
    port map
    (   '1',-- carry in is = 1
        a(a'high downto (PRTN_SIZ*NUM_SEG_REGS)),
        b(b'high downto (PRTN_SIZ*NUM_SEG_REGS)),
        s1,c1
    );
    -- select sum and carries for current segment
    c(c'high) <= c0 when c_reg(c_reg'high-1) = '0' else c1;
    s(s'high downto s'high-(EXTRA_BITS-1)) <=
        s0 when c_reg(c_reg'high-1) = '0' else s1;
end generate;

end RTL;
```

```
--------------------------------------------------------
--
--   cpa.vhd Carry Propagation Adder
--
--------------------------------------------------------
library IEEE;
use IEEE.std_logic_1164.all;
use IEEE.numeric_std.all;

entity cpa is
generic (SIZ: integer);
port
(
    -- inputs
    ci: in std_logic;
    a: in unsigned(SIZ-1 downto 0);
    b: in unsigned(SIZ-1 downto 0);
    -- outputs
    so: out unsigned(SIZ-1 downto 0);
    co: out std_logic
);
end cpa;

architecture RTL of cpa is

signal c: unsigned(SIZ-1 downto 0) := (others=>'0');

begin

-- created with full adders
gen_adr: for i in 0 to SIZ-1 generate

    gen_lsb: if i = 0 generate
        so(i) <= a(i) xor b(i) xor ci;
        c(i) <= (a(i) and ci) or (b(i) and ci) or (a(i) and b(i));
    end generate;

    gen_other: if i > 0 generate
        so(i) <= a(i) xor b(i) xor c(i-1);
        c(i) <= (a(i) and c(i-1)) or (b(i) and c(i-1)) or (a(i) and b(i));
    end generate;

end generate;
```

```
co <= c(c'high);

end RTL;
```

```
--------------------------------------------------------------
-- c_gen_sm.vhd (C generation state machine, which is the partial
-- result).
--
-- On-the-fly circuit is used to create cpr (positive result),
-- cmr (result minus 1 lsb), and cr (current partial result). All
-- three are fractional values with the sign bit to the left of
-- the binary point.
--------------------------------------------------------------
library IEEE;
use IEEE.std_logic_1164.all;
use IEEE.numeric_std.all;


entity c_gen_sm is
generic (C_REG_SIZ: integer := 64; INTERATIONS: integer);
port
(
    clk: in std_logic; -- system clock
    rst: in std_logic;  -- system reset
    -- inputs
    start: in std_logic;
    \qj+1\: in signed(1 downto 0);
    -- outputs
    cmplt: out std_logic;
    cr: out unsigned(C_REG_SIZ-1 downto 0);
    cpr: inout unsigned(C_REG_SIZ-1 downto 0);
    cmr: inout unsigned(C_REG_SIZ-1 downto 0)
);
end c_gen_sm;

architecture RTL of c_gen_sm is

signal sr: unsigned(C_REG_SIZ-1 downto 0) := (others=>'0');
signal j: natural range 1 to INTERATIONS-1 := 1;

signal cp_in: std_logic := '0';
signal cm_in: std_logic := '0';

signal cpr_sel: std_logic := '0';
signal cmr_sel: std_logic := '0';

signal cp: unsigned(C_REG_SIZ-1 downto 0) := (others=>'0');
signal cm: unsigned(C_REG_SIZ-1 downto 0) := (others=>'0');
```

```
signal cp_nth_bit: unsigned(C_REG_SIZ-1 downto 0) := (others=>'0');
signal cm_nth_bit: unsigned(C_REG_SIZ-1 downto 0) := (others=>'0');

attribute equivalent_register_removal: string;
attribute equivalent_register_removal of cr : signal is "no";
attribute equivalent_register_removal of cpr : signal is "no";
attribute equivalent_register_removal of cmr : signal is "no";

attribute keep:string;
attribute keep of cr :signal is "true";
attribute keep of cpr :signal is "true";
attribute keep of cmr :signal is "true";


----------------------
--  enumeration lists
----------------------
type sm_def is
(
    RESET,
    START_C_GEN,
    WAIT_ONE_CLK,
    GEN_NEXT_C,
    END_C_GEN
);
signal state: sm_def := RESET;

--------------------------------- module code ----------------------------
begin


process(rst,clk)
begin

    if(rst='1') then
        -- registers
        cr <= (others=>'0');
        cpr <= (others=>'0');
        cmr <= (others=>'0');
        -- misc
        j <= 1;
        sr <= (others=>'0');
        state <= RESET;
```

```
elsif rising_edge(clk) then

    --
    --   state machine body
    --
    case state is
        -- reset state
        when RESET =>
            state <= START_C_GEN;

        --
        --   C[j] generation
        --
        when START_C_GEN =>
            -- CR[1] = 0.1 binary, 2 to the -1 power
          cr <= (others=>'0');
            cr(cr'high-1)<= '1';
            cpr <= (others=>'0');
            cpr(cpr'high-1) <= '1';
            cmr <= (others=>'0');
            j <= 1;
            -- shift register indicating next bit position 0.1 binary,
            -- which will be right shifted in the next state.
            sr <= (others=>'0');
            sr(sr'high-1) <= '1';
            -- wait for the synchronous start
            if(start = '1') then
                state <= WAIT_ONE_CLK;
            end if;
    -- wait 1 clock for next qj+1
    when WAIT_ONE_CLK =>
            sr <= '0'&sr(sr'high downto 1);-- shift right
            state <= GEN_NEXT_C;
    -- register next C value
    when GEN_NEXT_C =>
            cr <= cp; -- cr is for external use
            cpr <= cp;
            cmr <= cm;
            --check for end of process
            if(j = INTERATIONS-1) then
                state <= END_C_GEN;
            else
                j <= j+1;
                state <= WAIT_ONE_CLK;
```

```
                    end if;
               -- wait for handshake to be released
               when END_C_GEN   =>
                   if(start = '0') then
                        state <= START_C_GEN;
                   end if;
               when others =>
                   state <= RESET;
           end case;

      end if;

end process;

cmplt <= '1' when (start = '1') and (state = END_C_GEN) else '0';


--
-- decode selection logic for both cp and cm, on-the-fly circuits
--
-- new input bit based on current qj+1
with \qj+1\ select -- 1 when +1 or -1, otherwise 0
    cp_in <= '1' when "01" | "11", '0' when others;
with \qj+1\ select -- 0 when +1 or -1, otherwise 1
    cm_in <= '0' when "01" | "11", '1' when others;

-- source selection for cpr and cmr based on current qj+1
with \qj+1\ select -- cpr and cp_in when 0 or +1, cmr and cp_in when -1
    cpr_sel <= '0' when "00" | "01", '1' when others;
with \qj+1\ select -- cmr and cm_in when 0 or -1, cpr and cm_in when +1
    cmr_sel <= '0' when "00" | "11", '1' when others;

-- use shift register to assign proper nth_bit position based on cp_in and cm_in
gen_cp_nth_bit: for i in cp_nth_bit'high downto 0 generate
    cp_nth_bit(i) <= sr(i) and cp_in;
end generate;
gen_cm_nth_bit: for i in cm_nth_bit'high downto 0 generate
    cm_nth_bit(i) <= sr(i) and cm_in;
end generate;

-- cpr and cmr inputs based on above selection
cp <= (cpr or cp_nth_bit) when cpr_sel = '0' else (cmr or cp_nth_bit);
cm <= (cmr or cm_nth_bit) when cmr_sel = '0' else (cpr or cm_nth_bit);

end RTL;
```

```
-------------------------------------------------------------
-- s_gen_sm.vhd (S generation state machine, which is the square of
-- the partial result).
--
-- Output for S is in carry-save form.
-------------------------------------------------------------
library IEEE;
use IEEE.std_logic_1164.all;
use IEEE.numeric_std.all;


entity s_gen_sm is
generic (C_REG_SIZ: integer ;S_REG_SIZ: integer; INTERATIONS: integer);
port
(
    clk: in std_logic; -- system clock
    rst: in std_logic;  -- system reset
    -- inputs
    start: in std_logic;
    \qj+1\: in signed(1 downto 0);
    cr: in unsigned(C_REG_SIZ-1 downto 0);
    -- squared partial result, 2C[j]
    cmplt: out std_logic;
    ssr: out unsigned(S_REG_SIZ-1 downto 0);
    scr: out unsigned(S_REG_SIZ-1 downto 0)
);
end s_gen_sm;

architecture RTL of s_gen_sm is


--------------
-- functions
--------------
-- d size can be up to the same size as S_REG_SIZ
function barrel_shift_right(d: unsigned;n: natural)  return unsigned is
variable input: unsigned(S_REG_SIZ-1 downto 0) := (others=>'0');
variable output: unsigned(S_REG_SIZ-1 downto 0) := (others=>'0');
begin

    -- place data in the upper region of input then shift
    input(input'high downto input'high-d'length+1) := d;
    output := shift_right(input,n);
    return(output);
```

```vhdl
end function;

----------------
-- components
----------------
component csa_4to2 is
generic (SIZ: integer);
port
(
    -- carry inputs
    cin1: in std_logic;
    cin2: in std_logic;
    -- operand inputs
    w: in unsigned(SIZ-1 downto 0);
    x: in unsigned(SIZ-1 downto 0);
    y: in unsigned(SIZ-1 downto 0);
    z: in unsigned(SIZ-1 downto 0);
    -- outputs (c0,s0/c1,s1/...)
    s: out unsigned(SIZ-1 downto 0);
    c: out unsigned(SIZ-1 downto 0);
    cout: out std_logic
);
end component;


----------------------
--   declared signals
----------------------
signal nth_digit: unsigned(S_REG_SIZ-1 downto 0) := (others=>'0');
signal j: natural range 1 to INTERATIONS-1 := 1;

signal ss: unsigned(S_REG_SIZ-1 downto 0) := (others=>'0');
signal sc: unsigned(S_REG_SIZ-1 downto 0) := (others=>'0');
-- i is for internal use
signal ssri: unsigned(S_REG_SIZ-1 downto 0) := (others=>'0');
signal scri: unsigned(S_REG_SIZ-1 downto 0) := (others=>'0');

signal c_for_sr: unsigned(S_REG_SIZ-1 downto 0) := (others=>'0');

-- other adder input
signal w: unsigned(S_REG_SIZ-1 downto 0) := (others=>'0');
signal z: unsigned(S_REG_SIZ-1 downto 0) := (others=>'0');
signal cin1: std_logic := '0';

-- duplication reduces fanout loading
```

```vhdl
attribute equivalent_register_removal: string;
attribute equivalent_register_removal of ssr : signal is "no";
attribute equivalent_register_removal of ssri : signal is "no";
attribute equivalent_register_removal of scr : signal is "no";
attribute equivalent_register_removal of scri : signal is "no";

attribute keep:string;
attribute keep of ssr :signal is "true";
attribute keep of ssri :signal is "true";
attribute keep of scr :signal is "true";
attribute keep of scri :signal is "true";


----------------------
--   enumeration lists
----------------------
type sm_def is
(
    RESET,
    START_S_GEN,
    GEN_C_FOR_S_ADDER,
    GEN_NEXT_S,
    END_S_GEN
);
signal state: sm_def := RESET;
---------------------------------- module code ----------------------------
begin

process(rst,clk)
begin

    if(rst='1') then
        -- registers
        ssr <= (others=>'0');
        scr <= (others=>'0');
        ssri <= (others=>'0');
        scri <= (others=>'0');
        c_for_sr <= (others=>'0');
        -- misc
        j <= 1;
        nth_digit <= (others=>'0');
        state <= RESET;

    elsif rising_edge(clk) then
        --
```

```
--  state machine body
--
case state is
    -- reset state
    when RESET =>
        state <= START_S_GEN;
    --
    --  S[j] generation body
    --
    when START_S_GEN =>
        -- SSR[1] = 0.01 binary, or 2 to the -2 power
        ssr <= (others=>'0');
        ssr(ssr'high-2) <= '1';
        ssri <= (others=>'0');
        ssri(ssr'high-2) <= '1';
        scr <= (others=>'0');
        scri <= (others=>'0');
        j <= 1;
        -- 0.0001 binary
        nth_digit <= (others=>'0');
        nth_digit(nth_digit'high-4) <= '1';
        -- wait for the synchronous start
        if(start = '1') then
            state <= GEN_C_FOR_S_ADDER;
        end if;
    -- generate C input to S adder
    when GEN_C_FOR_S_ADDER =>
        -- same as multiplying C by a single fractional digit
        c_for_sr <= barrel_shift_right(cr,j);
        state <= GEN_NEXT_S;
    -- generate next S value with other inputs
    when GEN_NEXT_S =>
        -- update S registers
        ssr <= ss;
        ssri <= ss;
        scr <= sc;
        scri <= sc;
        -- update nth digit value (right shift by 2)
        nth_digit <=   "00"&nth_digit(nth_digit'high downto 2);
        -- check for end of process
        if(j = INTERATIONS-1) then
            state <= END_S_GEN;
        else
            j <= j+1;
```

```
                state <= GEN_C_FOR_S_ADDER;
            end if;
        -- wait for handshake to be released
        when END_S_GEN =>
            if(start = '0') then
                state <= START_S_GEN;
            end if;
        when others =>
            state <= RESET;
    end case;

end if;

end process;

cmplt <= '1' when (start = '1') and (state = END_S_GEN) else '0';


--
--   other adder inputs
--
-- determines whether c_for_s is positive, negative, or zero
with \qj+1\ select
w <= c_for_sr when "01", not(c_for_sr) when "11",(others=>'0') when others;
cin1 <= '1' when (\qj+1\ = "11") else '0';
-- determines whether the nth digit is positive or zero
z <= nth_digit when (\qj+1\ /= 0) else (others=>'0');


--
--   adder for next S value in carry-save form
--
s_adder_mod: csa_4to2
generic map(SIZ => S_REG_SIZ)
port map
(
    -- carry inputs
    cin1 => cin1,
    cin2 => '0',
    -- operand inputs
    w => w,
    x => scri,
    y => ssri,
    z => z,
    -- outputs
    s => ss,
```

```
    c => sc,
    cout => open
);

end RTL;
```

```
------------------------------------------------------------
--
--   csa_4to2.vhd Carry Save Adder (2 levels of full adders)
--
--   Four operand adder with carry in and out. Cin is passed
--   c(0) to coincide with s(0).
--
------------------------------------------------------------
library IEEE;
use IEEE.std_logic_1164.all;
use IEEE.numeric_std.all;

entity csa_4to2 is
generic (SIZ: integer);
port
(
    -- carry inputs
    cin1: in std_logic; -- input to second csa level
    cin2: in std_logic; -- passes to c(0)
    -- operand inputs
    w: in unsigned(SIZ-1 downto 0);
    x: in unsigned(SIZ-1 downto 0);
    y: in unsigned(SIZ-1 downto 0);
    z: in unsigned(SIZ-1 downto 0);
    -- outputs (c0,s0/c1,s1/...)
    s: out unsigned(SIZ-1 downto 0);
    c: out unsigned(SIZ-1 downto 0);
    --
    cout: out std_logic
);
end csa_4to2;

architecture RTL of csa_4to2 is

signal sx1: unsigned(SIZ-1 downto 0) := (others=>'0');
signal cx1: unsigned(SIZ downto 1) := (others=>'0');
signal cx2: unsigned(SIZ downto 1) := (others=>'0');

begin

--
--   carry save adder level 1
--
gen_csa_1: for i in 0 to SIZ-1 generate
```

```
begin

    -- compute sum out
    sx1(i) <= (y(i) xor z(i)) xor x(i);
    -- compute carry out
    cx1(i+1) <= (y(i) and x(i)) or (z(i) and x(i)) or (y(i) and z(i));

end generate;


--
--  carry save adder level 2
--
gen_csa_2: for i in 0 to SIZ-1 generate
begin

    -- compute sum out
    gen_csa_sum_a:if(i = 0) generate
        s(0) <= sx1(0) xor w(0) xor cin1;
    end generate;
    gen_csa_sum_b:if(i > 0) generate
        s(i) <= sx1(i) xor w(i) xor cx1(i);
    end generate;

    -- compute carry out
    gen_csa_carry_a:if(i = 0) generate
        cx2(i+1) <= (sx1(0) and w(0)) or (sx1(0) and cin1) or (w(0) and cin1);
    end generate;
    gen_csa_carry_b:if(i > 0) generate
        cx2(i+1) <= (sx1(i) and w(i)) or (sx1(i) and cx1(i)) or (w(i) and cx1(i));
    end generate;

end generate;

-- format carry bits
c(0) <= cin2;
c(c'high downto 1) <= cx2(cx2'high-1 downto 1);
cout <= cx2(cx2'high);


end RTL;
```

```
--------------------------------------------------------------
-- w_gen_sm.vhd (W generation state machine, which is the current
-- residual).
--
-- Output for W is in carry-save form.
--------------------------------------------------------------
library IEEE;
use IEEE.std_logic_1164.all;
use IEEE.numeric_std.all;

entity w_gen_sm is
generic (C_REG_SIZ: integer; S_REG_SIZ: integer; W_REG_SIZ: integer;
INTERATIONS: integer);
port
(
    clk: in std_logic; -- system clock
    rst: in std_logic;  -- system reset
    -- inputs
    start: in std_logic;
    w_init: in unsigned(W_REG_SIZ-1 downto 0);
    cr: in unsigned(C_REG_SIZ-1 downto 0);
    scr: in unsigned(S_REG_SIZ-1 downto 0);
    ssr: in unsigned(S_REG_SIZ-1 downto 0);
    -- current residual
    cmplt: out std_logic;
    wsr: inout unsigned(W_REG_SIZ-1 downto 0);
    wcr: inout unsigned(W_REG_SIZ-1 downto 0);
    -- next signed digit
    \c qj+1\: out signed(1 downto 0);
    \s qj+1\: out signed(1 downto 0)
);
end w_gen_sm;

architecture RTL of w_gen_sm is


--------------
-- functions
--------------
-- d size can be up to the same size as S_REG_SIZ
function barrel_shift_right(d: unsigned;n: natural)  return unsigned is
-- do not include the sign bit in shift
variable input: unsigned(S_REG_SIZ-2 downto 0) := (others=>'0');
variable output: unsigned(S_REG_SIZ-1 downto 0) := (others=>'0');
begin
```

```
    -- place data in the upper region of input then shift
    input(input'high downto input'high-d'length+2) := d(d'high-1 downto 0);
    output := '0'&shift_right(input,n);
    return(output);

end function;


---------------
-- components
---------------
-- 3 operand to two output adder
component csa_3to2 is
generic (SIZ: integer);
port
(
    -- carry input
    cin: in std_logic;
    -- operand inputs
    x: in unsigned(SIZ-1 downto 0);
    y: in unsigned(SIZ-1 downto 0);
    z: in unsigned(SIZ-1 downto 0);
    -- outputs (c0,s0/c1,s1/...)
    s: out unsigned(SIZ-1 downto 0);
    c: out unsigned(SIZ-1 downto 0);
    --
    cout: out std_logic
);
end component;

-- 6 operand to two output adder tree
component csa_6to2 is
generic (SIZ: integer);
port
(
    -- carry inputs
    cin1: in std_logic;
    cin2: in std_logic;
    cin3: in std_logic;
    cin4: in std_logic;
    -- operand input
    x1: in unsigned(SIZ-1 downto 0);
    x2: in unsigned(SIZ-1 downto 0);
    x3: in unsigned(SIZ-1 downto 0);
```

```
    x4: in unsigned(SIZ-1 downto 0);
    x5: in unsigned(SIZ-1 downto 0);
    x6: in unsigned(SIZ-1 downto 0);
    -- outputs
    s: out unsigned(SIZ-1 downto 0);
    c: out unsigned(SIZ-1 downto 0);
    --
    cout: out std_logic
);
end component;



----------------------
-- declared signals
----------------------
signal j: natural range 1 to INTERATIONS-1 := 1;
signal \j+1\: natural range 2 to INTERATIONS := 2;

signal \early qj+1\: signed(1 downto 0) := (others=>'0');
signal \qj+1\: signed(1 downto 0) := (others=>'0');


-- size of S is used because of the lower number of bits is twice
-- that of C in order to be combined with W.
signal c_for_w: unsigned(S_REG_SIZ-1 downto 0) := (others=>'0');
-- C data uses W size + 1 integer bits to account for 3 x C and fit into adder tree
signal cx1: unsigned((W_REG_SIZ+1)-1 downto 0) := (others=>'0');
signal cy1: unsigned((W_REG_SIZ+1)-1 downto 0) := (others=>'0');
signal cz1: unsigned((W_REG_SIZ+1)-1 downto 0) := (others=>'0');
signal csx3: unsigned((W_REG_SIZ+1)-1 downto 0) := (others=>'0');
signal ccx3: unsigned((W_REG_SIZ+1)-1 downto 0) := (others=>'0');
signal csx3r: unsigned((W_REG_SIZ+1)-1 downto 0) := (others=>'0');
signal ccx3r: unsigned((W_REG_SIZ+1)-1 downto 0) := (others=>'0');
-- S uses S + 2 to account for times 3 plus making it equal to W
signal sx123: unsigned((S_REG_SIZ+2)-1 downto 0) := (others=>'0');
signal sx456: unsigned((S_REG_SIZ+2)-1 downto 0) := (others=>'0');
signal ssx3: unsigned((S_REG_SIZ+2)-1 downto 0) := (others=>'0');
signal scx3: unsigned((S_REG_SIZ+2)-1 downto 0) := (others=>'0');
-- S uses S + 3 to account for times 3 plus making it equal to W+1 (adder tree)
signal ssx3r_in: unsigned((S_REG_SIZ+3)-1 downto 0) := (others=>'0');
signal scx3r_in: unsigned((S_REG_SIZ+3)-1 downto 0) := (others=>'0');
signal ssx3r: unsigned((S_REG_SIZ+3)-1 downto 0) := (others=>'0');
signal scx3r: unsigned((S_REG_SIZ+3)-1 downto 0) := (others=>'0');
```

```
-- W registers
signal nth_digit: unsigned((W_REG_SIZ+1)-1 downto 0) := (others=>'0');

-- pre W adder x 2 W, requires extra integer bit
signal ws: unsigned((W_REG_SIZ+1)-1 downto 0) := (others=>'0');
signal wc: unsigned((W_REG_SIZ+1)-1 downto 0) := (others=>'0');
signal wx1: unsigned((W_REG_SIZ+1)-1 downto 0) := (others=>'0');
signal wy1: unsigned((W_REG_SIZ+1)-1 downto 0) := (others=>'0');
signal wz1: unsigned((W_REG_SIZ+1)-1 downto 0) := (others=>'0');
signal wz1_cin: std_logic := '0';
signal ws_wnth: unsigned((W_REG_SIZ+1)-1 downto 0) := (others=>'0');
signal wc_wnth: unsigned((W_REG_SIZ+1)-1 downto 0) := (others=>'0');
signal ws_wnthr: unsigned((W_REG_SIZ+1)-1 downto 0) := (others=>'0');
signal wc_wnthr: unsigned((W_REG_SIZ+1)-1 downto 0) := (others=>'0');

-- W adder tree inputs have 1 more integer bit that W + 1 integer bit
signal x1: unsigned((W_REG_SIZ+1)-1 downto 0) := (others=>'0');
signal x2: unsigned((W_REG_SIZ+1)-1 downto 0) := (others=>'0');
signal x3: unsigned((W_REG_SIZ+1)-1 downto 0) := (others=>'0');
signal x4: unsigned((W_REG_SIZ+1)-1 downto 0) := (others=>'0');


-- adder tree carry inputs
signal w_cin1: std_logic := '0';
signal w_cin2: std_logic := '0';

signal w_estimate: unsigned(4 downto 0) := (others=>'0');

attribute equivalent_register_removal: string;
attribute equivalent_register_removal of \qj+1\ : signal is "no";
attribute equivalent_register_removal of \c qj+1\ : signal is "no";
attribute equivalent_register_removal of \s qj+1\ : signal is "no";

attribute keep:string;
attribute keep of \qj+1\ :signal is "true";
attribute keep of \c qj+1\ :signal is "true";
attribute keep of \s qj+1\ :signal is "true";


-----------------------
--   enumeration lists
-----------------------

type sm_def is
(
```

```
    RESET,
    START_W_GEN,
    GEN_ADDER_INPUTS,
    GEN_NEXT_W,
    END_W_GEN
);
signal state: sm_def := RESET;

--------------------------------- module code ----------------------------
begin


process(rst,clk)
begin

    if(rst='1') then
        -- registers
        wsr <= (others=>'0');
        wcr <= (others=>'0');
        csx3r <= (others=>'0');
        ccx3r <= (others=>'0');
        ssx3r <= (others=>'0');
        scx3r <= (others=>'0');
        ws_wnthr <= (others=>'0');
        wc_wnthr <= (others=>'0');
        -- misc
        j <= 1;
        \j+1\ <= 2;
        \qj+1\ <= (others=>'0');
        \c qj+1\ <= (others=>'0');
        \s qj+1\ <= (others=>'0');
        nth_digit <= (others=>'0');

        state <= RESET;

    elsif rising_edge(clk) then

        --
        --   state machine body
        --
        case state is
            -- reset state
            when RESET =>
                state <= START_W_GEN;
```

```
--
--   W[j] generation body
--
when START_W_GEN =>
    -- W[1] = 2X - (2 to the -2 power)
    -- WSR[1] = operand X2
    -- WCR[1] = 111.11 binary, or 2 to the -2 power
    wsr <= w_init(w_init'high-1 downto 0)&'0';
    wcr <= (others=>'0');
    wcr(wcr'high downto wcr'high-4) <= "11111";
    j <= 1;
    \j+1\ <= 2;
    -- 0000.0001 binary
    nth_digit <= (others=>'0');
    nth_digit(nth_digit'high - 7) <= '1';
    -- initial qj+1
    \qj+1\ <= "00";
    \c qj+1\ <= "00";
    \s qj+1\ <= "00";
    -- wait for the synchronous start
    if(start = '1') then
        state <= GEN_ADDER_INPUTS;
    end if;
-- generate input to adder tree along with qj+1 for
-- external modules
when GEN_ADDER_INPUTS =>
-- post qj+1
    \qj+1\ <= \early qj+1\;
    \c qj+1\ <= \early qj+1\;
    \s qj+1\ <= \early qj+1\;
    -- register W adder tree with prepared C value
    csx3r <= csx3;
    ccx3r <= ccx3;
    -- register W adder tree with prepared S
    ssx3r <= ssx3r_in;
    scx3r <= scx3r_in;
    -- register W adder tree input with nth digit
    ws_wnthr <= ws_wnth;
    wc_wnthr <= wc_wnth;
    state <= GEN_NEXT_W;
-- generate next W value based on output from adder tree
when GEN_NEXT_W =>
    wsr <= ws(ws'high-1 downto 0); -- discard the msb
    wcr <= wc(wc'high-1 downto 0);
```

```
            -- update nth digit value (right shift by 2)
            nth_digit <=    "00"&nth_digit(nth_digit'high downto 2);
            -- check for end of process
            if(j = INTERATIONS-1) then
                state <= END_W_GEN;
            else
                j <= j+1;
                \j+1\ <= \j+1\ + 1;
                state <= GEN_ADDER_INPUTS;
            end if;
        -- wait for handshake to be released
        when END_W_GEN =>
            if(start = '0') then
                state <= START_W_GEN;
            end if;
        when others =>
            state <= RESET;
    end case;

  end if;

end process;

cmplt <= '1' when (start = '1') and (state = END_W_GEN) else '0';


------------------------
--
--   C preparation logic
--
-- same as multipling C by a single fractional digit
c_for_w <= barrel_shift_right(cr,\j+1\);
-- multiply C[j] by 3, adding times 2 and times 1 as partial products
-- equals times 3. inverting and adding 1 to each partial products
-- results in subtract of C from W. result is in carry save form and fits
-- into the final adder tree.
cx1 <= not("00"&c_for_w&'0'); -- times 2
cy1 <= not("000"&c_for_w); -- times 1
cz1 <= to_unsigned(1,cz1'length); -- add 1 to subtract

  -- W+1 size to fit directly into adder tree
c_csa_3to2_mod: csa_3to2
generic map (SIZ => (W_REG_SIZ+1))
port map
```

```
(
    -- carry input
    cin => '1', -- add 1 to subtract
    -- operand inputs
    x => cx1,
    y => cy1,
    z => cz1,
    -- outputs (c0,s0/c1,s1/...)
    s => csx3,
    c => ccx3,
    --
    cout => open
);
with \qj+1\ select -- subtract 3C or add zero
    x1 <= csx3r when "01" | "11", (others=>'0') when others;
with \qj+1\ select
    x2 <= ccx3r when "01" | "11", (others=>'0') when others;


-----------------------
    --
    --   S preparation logic
    --
    -- multiply S[j] by 3, adding the sum and carry components 3 times
    -- each equals times 3. inverting and adding 1 to each partial products
    -- (main adder tree) results in subtract S from W. result is in
    -- carry save form.
    --
    -- sign extend sum and carry to W width
process(ssr,scr)
begin
    -- evaluate sum and carry bits to determine sign extension
    if(ssr(ssr'high) = '0' and scr(scr'high) = '0') then
        sx123 <= "00"&ssr;
        sx456 <= "00"&scr;
    elsif(ssr(ssr'high) = '1' and scr(scr'high) = '0') then
        sx123 <= "11"&ssr;
        sx456 <= "00"&scr;
    elsif(ssr(ssr'high) = '0' and scr(scr'high) = '1') then
        sx123 <= "00"&ssr;
        sx456 <= "11"&scr;
    else -- both are 1
        sx123 <= "11"&ssr;
        sx456 <= "00"&scr;
    end if;
```

```
end process;

-- module multiplies S times 3
csa_6to2_mod_for_s: csa_6to2
generic map (SIZ => S_REG_SIZ+2) -- width matches W
port map
(
    -- carry inputs
    cin1 => '0',
    cin2 => '0',
    cin3 => '0',
    cin4 => '0',
    -- operand input
    x1 => sx123, -- ssr x 3
    x2 => sx123,
    x3 => sx123,
    x4 => sx456, -- scr x 3
    x5 => sx456,
    x6 => sx456,
    -- outputs
    s => ssx3,
    c => scx3,
    --
    cout => open
);
-- extend sign adder output to fit adder tree for x3 and x4
process(ssx3,scx3)
begin
    -- evaluate sum and carry bits to determine sign extension
    if(ssx3(ssx3'high) = '0' and scx3(scx3'high) = '0') then
        ssx3r_in <= "0"&ssx3;
        scx3r_in <= "0"&scx3;
    elsif(ssx3(ssx3'high) = '1' and scx3(scx3'high) = '0') then
        ssx3r_in <= "1"&ssx3;
        scx3r_in <= "0"&scx3;
    elsif(ssx3(ssx3'high) = '0' and scx3(scx3'high) = '1') then
        ssx3r_in <= "0"&ssx3;
        scx3r_in <= "1"&scx3;
    else -- both are 1
        ssx3r_in <= "1"&ssx3;
        scx3r_in <= "0"&scx3;
    end if;
end process;
```

```
-- determine if S x 3 is added, subtracted, or zero'ed
with \qj+1\ select
    x3 <= not(ssx3r) when "01", -- subtracting S from W
              (ssx3r)  when "11", -- adding S to W
              (others=>'0') when others; -- add zero to W instead
with \qj+1\ select
    x4 <= not(scx3r)  when "01", -- subtracting S from W
              (scx3r) when "11", -- adding S to W
              (others=>'0') when others; -- add zero to W instead


-- if the above is inverted set corresponding carry into main adder tree
with \qj+1\ select
    w_cin1 <= '1' when "01", '0' when others; -- subtract
with \qj+1\ select
    w_cin2 <= '1' when "01", '0' when others; -- subtract


----------------------------------------
--
--   W preparation and generation logic
--
-- computed signed digit for next iteration using using upper 5-bits of both
-- sum and carry, mimicing x2
--
w_estimate <= wsr(wsr'high downto wsr'high-4) +  wcr(wcr'high downto wcr'high-4);
process(w_estimate)
begin

    -- +1 if greater than 0
    if(signed(w_estimate) > 0) then
        \early qj+1\ <= "01";
    -- 0 if between -1/2 and zero
    elsif(w_estimate = "11111" or w_estimate = 0) then
        \early qj+1\ <= "00";
    -- -1 if equal to or less than -1
    else
        \early qj+1\ <= "11";
    end if;

end process;


--
--   combine W and nth digit into 3 to 2 adder in carry-save form
--
wx1 <= wsr&'0'; -- times 2
```

```
wy1 <= wcr&'0';
-- add or subtract nth digit
with \early qj+1\ select -- an add means subtract
    wz1 <= not(nth_digit) when "01", nth_digit when "11", (others=>'0') when others;
with \early qj+1\ select
    wz1_cin <= '1' when "01", '0' when others;

w_csa_3to2_mod: csa_3to2
generic map (SIZ => W_REG_SIZ+1)
port map
(
    -- carry input
    cin => wz1_cin,
    -- operand inputs
    x => wx1,
    y => wy1,
    z => wz1,
    -- outputs (c0,s0/c1,s1/...)
    s => ws_wnth, -- pass to maain adder tree
    c => wc_wnth,
    --
    cout => open
);


--
-- 6 input adder tree to generate W[j+1] in carry-save form
--
csa_6to2_mod_for_w: csa_6to2
generic map (SIZ => W_REG_SIZ+1)
port map
(
    -- carry inputs
    cin1 => w_cin1,
    cin2 => w_cin2,
    cin3 => '0',
    cin4 => '0',
    -- operand input
    x1 => x1,  -- prepared 3C[j] sum and carry
    x2 => x2,
    x3 => x3, -- prepared 3S[j] sum and carry
    x4 => x4,
    x5 => ws_wnthr, -- prepared W with nth digit
    x6 => wc_wnthr,
    -- outputs
```

```
    s => ws,
    c => wc,
    --
    cout => open
);

end RTL;
```

```
---------------------------------------------------------
--
--   csa_3to2.vhd Carry Save Adder (1 level of full adders)
--
--   Three operand adder with carry in and out. Cin is passed
--   c(0) to coincide with s(0).
--
---------------------------------------------------------
library IEEE;
use IEEE.std_logic_1164.all;
use IEEE.numeric_std.all;

entity csa_3to2 is
generic (SIZ: integer);
port
(
    -- carry input
    cin: in std_logic;
    -- operand inputs
    x: in unsigned(SIZ-1 downto 0);
    y: in unsigned(SIZ-1 downto 0);
    z: in unsigned(SIZ-1 downto 0);
    -- outputs (c0,s0/c1,s1/...)
    s: out unsigned(SIZ-1 downto 0);
    c: out unsigned(SIZ-1 downto 0);
    --
    cout: out std_logic
);
end csa_3to2;

architecture RTL of csa_3to2 is

signal cx: unsigned(SIZ downto 1) := (others=>'0');

begin

--
--   carry save adder
--
gen_csa: for i in 0 to SIZ-1 generate
begin

    -- compute sum out
    s(i) <= (y(i) xor z(i)) xor x(i);
```

```vhdl
      -- compute carry out
      cx(i+1) <= (y(i) and x(i)) or (z(i) and x(i)) or (y(i) and z(i));

end generate;

-- format carry bits
c(0) <= cin;
c(c'high downto 1) <= cx(cx'high-1 downto 1);
cout <= cx(cx'high);

end RTL;
```

```
-----------------------------------------------------------
--
--    csa_6to2.vhd Carry Save Adder (3 levels of full adders)
--
--    Six operand adder with carry in and out. There are a
--    total of 4 carry inputs, one for each individual carry
--    save adder. Each carry in coincides with the s(0) out
--    of each stage.
--
--    The supports two's complement conversion. Inverting one
--    the operands and setting the cin true. Up to four such
--    conversions are possible.
--
--        Note:x1, x2, x3 have only 2 levels.
--
-----------------------------------------------------------
library IEEE;
use IEEE.std_logic_1164.all;
use IEEE.numeric_std.all;

entity csa_6to2 is
generic (SIZ: integer);
port
(
    -- carry inputs (number is not related to operand number)
    cin1: in std_logic;
    cin2: in std_logic;
    cin3: in std_logic;
    cin4: in std_logic;
    -- operand inputs
    x1: in unsigned(SIZ-1 downto 0);
    x2: in unsigned(SIZ-1 downto 0);
    x3: in unsigned(SIZ-1 downto 0);
    x4: in unsigned(SIZ-1 downto 0);
    x5: in unsigned(SIZ-1 downto 0);
    x6: in unsigned(SIZ-1 downto 0);
    -- outputs
    s: out unsigned(SIZ-1 downto 0);
    c: out unsigned(SIZ-1 downto 0);
    --
    cout: out std_logic
);
end csa_6to2;
```

```
architecture RTL of csa_6to2 is

signal sx1a,sx1b: unsigned(SIZ-1 downto 0) := (others=>'0');
signal sx2: unsigned(SIZ-1 downto 0) := (others=>'0');

signal cx1a,cx1b: unsigned(SIZ-1 downto 1) := (others=>'0');
signal cx2: unsigned(SIZ-1 downto 1) := (others=>'0');
signal cx3: unsigned(SIZ downto 1) := (others=>'0');


begin

--
--   carry save adder level 1
--
gen_csa_1: for i in 0 to SIZ-1 generate
begin

    -- compute sum out
    sx1a(i) <= (x1(i) xor x2(i)) xor x3(i);
    sx1b(i) <= (x4(i) xor x5(i)) xor x6(i);

    -- compute carry out
    gen_carry_lev1:if(i < SIZ-1) generate
        cx1a(i+1) <= (x1(i) and x2(i)) or (x1(i) and x3(i)) or (x2(i) and x3(i));
        cx1b(i+1) <= (x4(i) and x5(i)) or (x4(i) and x6(i)) or (x5(i) and x6(i));
    end generate;

end generate;

--
--   carry save adder level 2
--
gen_csa_2: for i in 0 to SIZ-1 generate
begin

    -- compute sum out
    gen_sum_lev2_a:if(i = 0) generate
        sx2(0) <= (sx1a(0) xor cin1) xor sx1b(0);
    end generate;
    gen_sum_lev2_b:if(i > 0) generate
        sx2(i) <= (sx1a(i) xor cx1b(i)) xor sx1b(i);
    end generate;
```

```
    -- compute carry out
    gen_carry_lev2_a:if(i = 0) generate
        cx2(i+1) <= (sx1a(0) and cin1) or (sx1a(0) and sx1b(0)) or (sx1b(0) and
cin1);
    end generate;
    gen_carry_lev2_b:if(i > 0 and i < SIZ-1) generate
        cx2(i+1) <= (sx1a(i) and cx1b(i)) or (sx1a(i) and sx1b(i)) or (cx1b(i) and
sx1b(i));
    end generate;

end generate;


--
--   carry save adder level 3
--
gen_csa_3: for i in 0 to SIZ-1 generate
begin

    -- compute sum out
    gen_sum_lev3_a:if(i = 0) generate
        s(0) <= (sx2(0) xor cin2) xor cin3;
    end generate;
    gen_sum_lev3_b:if(i > 0) generate
        s(i) <= (cx1a(i) xor cx2(i)) xor sx2(i);
    end generate;

    -- compute carry out
    gen_carry_lev3_a:if(i = 0) generate
        cx3(i+1) <= (cin2 and cin3) or (cin2 and sx2(0)) or (cin3 and sx2(0));
    end generate;
    gen_carry_lev3_b:if(i > 0) generate
        cx3(i+1) <= (cx1a(i) and cx2(i)) or (cx1a(i) and sx2(i)) or (cx2(i) and sx2(i));
    end generate;

end generate;

-- format carry bits
c(0) <= cin4;
c(c'high downto 1) <= cx3(cx3'high-1 downto 1);
cout <= cx3(cx3'high);


end RTL;
```

8   Nth Root Algorithm (Newton-Raphson)

Nth Root simply means the algorithm can solve for a range of orders, in this case 2 to 5, and can do so on the fly. The index is an input parameter along with the radicand itself.

$$\sqrt[n]{x} \text{ where index } n = 2, 3, 4, \text{ or } 5$$

The algorithm of choice is a derivation of the renowned Newton-Raphson method. It is iterative, it has a quadratic rate of convergence, and is surprisingly simple. However, it does require a starting guess. This last point has inspired a great deal of effort by others to quicken the initial steps but they are not employed here. Instead the simplest of means are used.

The design is fixed-point, with scalable integer and fractional portions (qualified), and a parameter setting the number of significant bits governing error of convergence. All negative radicands are first converted to positive operands as required by the algorithm and back to negative if required by the index value. An imaginary only bit is used in some cases.

> Note: *The Newton-Raphson algorithm is more suitable for floating-point because the size of intermediate terms can be very large. This is accounted for by pre-scaling of the radicand.*

External multipliers, dividers, and power functions are required to augment this design. All adders needed are implemented within the design. A section that follows provides recommendations for external resources.

8.1   Base Algorithm

Below is the mathematical expression where n represents the index (order) of the root being extracted, R represents the input radicand, i the current iteration, and X the current root value at that iteration. X_i and X_{i+1} are the current intermediate root value and the next, respectively. The first X_i (X_0) is the initial guess.

$$X_{i+1} = 1/n \cdot [\, ((n-1) \cdot X_i) + (R/(X_i^{\,n-1})) \,]$$

After each iteration X_i and X_{i+1} are compared and if difference is below an acceptable value the loop is exited. This difference is a predetermined value set by the designer and is sometimes referred to as an error margin. Most convergent algorithms use this techique.

8.2   Operational Aspects

8.2.1   Sequence

Figure 23 illustrates the overall flow of the Nth-Root function. The mathematical expression is partitioned in an attempt to parallel process as many of the terms as possible. Steps are as follows:

- Compute initial guess as explained in section 8.2.3 *Initial Guess.*

- Compute terms $t1 = ((n - 1) \bullet X_i)$ and $t2 = (X_i^{\,n-1})$

- Compute term $t3 = (R/t2)$

- Compute sum $s1 = t1 + t3$

- Compute $X_{i+1} = s1 \bullet 1/n$

- Compare convergence difference between X_i and X_{i+1}

Note: *the above steps are overlapped to reduce the number of clocks. The last 5 bullets take five clocks and is considered one iteration.*

All negative radicands are first converted to positive ones, then scaled into a workable range. Then the initial guess is made creating X_0 which is the seed to the algorithm. Iterations compute intermediate terms and sums in multiple steps resulting in X_{i+1}. Convergence is then tested, and if it is within tolerance the algorithm is exited. Rounding is evaluated and the root is converted to its proper sign. Lastly the result is renormalized based on the initial prescaling.

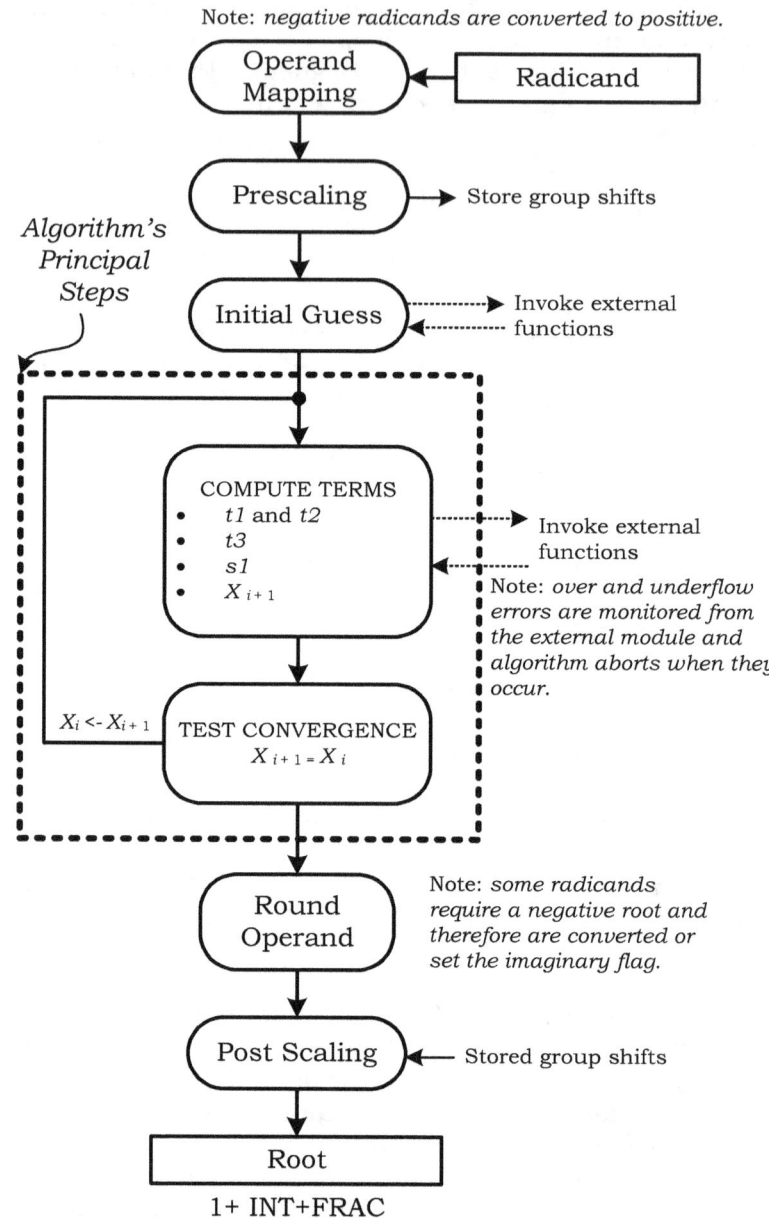

Figure 23

8.2.2 Operand Scaling

Term $t2$ is the power of X_i which can be exceedingly large with the higher indexes so must be limited in magnitude. The operand must be reduced by right shifting in groups of bits based on the index. Example an index of 5 requires 5-bit shifts, 4 requires 4-bit shifts, 3 requires 3-bits. Each group shift is recorded for post scaling or renormalizing the extracted root maintaining the 5:1, 4:1, and 3:1 ratio, by left shifting the number of group shifts.

   Note: *if there is no integer present either by format or content, no pre or post scaling is required.*

Group shifts continue until the largest possible integer for that index is reached as listed below. Because there has to be some internal integer content (which is fixed constant) for computing terms the largest values listed below are approximately 4 times smaller than they have to be. They serve as recommended ranges based on a Q31.32 format while providing some extra range for all indexes. The initial guess also reduces the size by making the initial result smaller.

Index	Largest value	Number of bits to evaluate
2	$4{,}294{,}967{,}295^{\,2\text{-}1}$	31-bits (requires no prescaling)
3	$65{,}535^{\,3\text{-}1}$	16-bits
4	$1023^{\,4\text{-}1}$	10-bits
5	$255^{\,5\text{-}1}$	8-bits

Side effects

Right shifting during prescaling will discard lower bits of the operand, but this is a trade-off for accommodating operands with large integer content. Its effect is less for lower indexes and nonexistent for square roots. Rounding is done before the root is left shifted.

As with having a selectable index on the fly, prescaling is also selectable as well. Large operands will however cause overflows to occur.

Other Considerations

This does not consider prescaling, but after a number of iterations term $t2$ can become very small compared to R resulting in an overflow when computing term *t3*. Conversely, if the input radicand is an extremely small fractional number the power function will underflow, especially with the higher indexes.

8.2.3 Initial Guess

The initial guess X_0 gives the algorithm a starting point. The closer it is to the resultant root the faster convergence will occur. As stated earlier a "great deal of effort by others have been made" to accomplish this. However, these guess-algorithms are as complicated or more complicated than the Newton-Raphson itself. The Bisection method was considered but it requires two initial guesses where one value has to be greater and one less than the resulting root.

> Note: *The initial guess here simply multiplies or divides the prescaled input radicand by its index (root order), then Newton-Raphson does the rest.*

An external multiplier/divider is used for all for root orders (not a simple shift) and depending on whether the operand is fractional only, they are treated different. Below fixed-point format and corresponding qualifiers.

Radicand divided by index

integer only format
integer and fractional format with some integer content

Radicand multiplied by index (limited to less than 1)

fractional only format
integer and fractional format with fractional content only

8.2.4 Sign versus Imaginary

The algorithm can only operate on positive numbers. Input radicands can be any sign regardless of the index but all output roots will follow rules as spelled out in section 2.5 *Index and Sign of Radicand*. Roots that cannot be a negative value will use the imaginary output to represent a negative result while the root itself is positive.

8.2.5 External Resources

Having external multipliers and dividers allows the designer to scale the performance of the design by the capability of the target FPGA device. While multiplier and divider modules covered in this book series can be used, it is highly recommended that DSP hard IP be employed because they can execute in 1 to several clocks. Most DSP circuits can be configured as either multipliers or dividers in a fixed-point format -- having some limitations to the range of integer/fractional bits. Designing one's own by using partial product and pipelined multipliers is also an effective solution. One goal of the *Nth Root* algorithm is to have medium to high performance and taking advantage of dedicated DSP silicon or embedded multipliers is the way to reach it.

The design requires the following external resources:

- One multiplier
- One powers (can also use a second multiplier for this)
- One divider

> Note: *rounding is done within the external modules themselves.*

Some functions are used concurrently in order to parallel intermediate steps in the overall process.

> Note: *the NthRootPackage.vhd file contains constants for operand size which can be used by the external module.*

8.2.6 Convergence

Convergence occurs when the current intermediate root and the next one, X_i and X_{i+1}, become close to equal or equal. How close to is based on a user settable parameter and defaults to 1 LSB. This takes a number of iterations which is largely determined by the length of the operand and how close the initial guess is to the resulting root. The *Nth Root* design default will include all integer and fractional bits plus extra bits for rounding for testing convergence.

> Note: *in addition to comparing X_i and X_{i+1}, the last difference is also monitored. In the event they are equal and differ only by their*

sign, this is also considers convergence. Otherwise the algorithm would oscillate around a central value.

The number of fractional bits can be limited by a parameter. The typical formula is 32-bits/3.322 = ~9 decimal places. The value 3.322 is the ratio between binary and decimal digits. Six to 8 equivalent decimal places are usually considered sufficient.

Note: *Rounding bits influence when a result converges.*

Review intro in chapter 5 as well as sections 5.3 through 5.6 to see how rounding is addressed in this book series. The previous section on convergence should be considered as well.

Because external functions are used, those module contain their own rounding and should be properly implemented.

Because of a compounding effect of external modules and the algorithm itself, it has been observed that the odd indexes like 3 and 5 result in a smaller root so have to be treated differently.

Rounding Notes

Rounding bits include guard and round with two additional bits for sticky (since there is no residual for sticky).

Even index rounding GRS > "1000" or (GRS = "1000" and LSB = '1')

Odd index rounding GSR /= 0

Warning: *Large fractional results are not prevented from rounding to 1.0 when enabled. The reader may want to account for this.*

8.2.7 Conclusion

Performance

All internal adders in the design are propagation adders which limit the overall clock speed. Sign conversion and the initial guess take 3 clocks plus the number of clocks the external multiplier and divider take. If DSP IP solutions are used it generally takes 1, partial product pipelining takes a few more. The algorithm itself takes 5 clocks per iteration in addition to the number of clocks of the external modules. Final steps require a few more clocks for rounding and the final sign conversion. Then there are several clocks for pre and post scaling due to shifting. Performance is highly data dependent.

Optional Improvements

- Reduce the number of bits required for convergence. The default requires all bits, including the added guard, round, and sticky bits.
- Square roots terms $t1$ and $t2$ reduce to X_i and require no multiplication or power operation. Saving a clock every iteration.
- Using a shifts for the initial guess when the index is either 2 or 4.
- Add duplication to internal registers to reduce fan-out.

Example Design

The example state machine provided illustrates the *Nth Root*.

- Format is scalable fixed-point Qm.n, integer and fractional bit lengths are parameters, defaulting to Q31.32, with a range to Q63.64. Other sizes may be used but the formentioned considerations must be accounted for in particular sizing of internal data paths and magnitude of the terms.
- Signed two's complement numbers are supported for the input radicand. All conversions are handled and managed internally by the state machine. The index and the input radicand sign determine the sign of the root or whether the imaginary bit is used.
- Rounding is dynamic, being enabled or disabled during each operation. Defaults to round-to-nearest-even and away from 0 depending on the index.
- The index is dynamic and scaling ban be enabled or disabled during each operation.

  Note: *parameters constant are contained in the nthRootPackage.vhd package file. This allows external modules to reference operand size.*

As configured for Q31.32, test builds were run using Xilinx ISE.

The following performance was obtained with corresponding part. While a Spartan could be used higher performance parts are recommended.

Xilinx Virtex XC5vlx110 greater than 213MHz

Note: *The part above was chosen for the package size because no internal multipliers are used, they merely use the IO pins.. When implemented it is expected that internal multipliers, dividers and power functions are internal to the FPGA.*

```
--------------------------------------------------------------------
--   nthRoot.vhd design (Newton-Raphson)
--------------------------------------------------------------------
library IEEE;
use IEEE.std_logic_1164.all;
use IEEE.numeric_std.all;
--use IEEE.math_real.all;
-- user packages
use work.nthRootPackage.all;

entity nthRoot is
--
--   Qmn fixed point format is used.
--
--        sign        binary point
--         |           |
--   format <s>(integer bits).(fractional bits)
--            _____/ _____/
--           INT_SIZ       FRAC_SIZ
--
-- Set in nthRootPackage.vhd file.notes about default setting.
--
port
(
    clk,rst: in std_logic; -- system clock and reset (synchronous)
    -- inputs
    n: in natural range 2 to 5; -- root order
    scale_en: in std_logic; -- pre/post scale enable
    rnd_en: in std_logic; -- enable rounding for current cycle
    extr: in std_logic; -- initiate root extraction
    rcd: in signed((1+INT_SIZ+FRAC_SIZ)-1 downto 0); -- radicand
    -- interface to all external modules
    -- interface to external multiplier
    startm: out std_logic; -- start multiplication
    multiplier: out signed(OP_SIZ-1 downto 0);
    multiplicand: out signed(OP_SIZ-1 downto 0);
    cmpltm: in std_logic; -- multiplication complete
    ovrflwm: in std_logic; -- overflow error
    product: in signed(OP_SIZ-1 downto 0);
    -- interface to external powers
    startp: out std_logic; -- start power function
    power: out natural range 1 to 4;
    base: out signed(OP_SIZ-1 downto 0);
    cmpltp: in std_logic; -- power complete
```

```vhdl
    ovrflwp: in std_logic; -- overflow error
    udrflwp: in std_logic; -- underflow error
    result: in signed(OP_SIZ-1 downto 0);
    -- interface to external divider
    startd: out std_logic; -- start division
    divisor: out signed(OP_SIZ-1 downto 0);
    dividend: out signed(OP_SIZ-1 downto 0);
    cmpltd: in std_logic; -- division complete
    ovrflwd: in std_logic; -- overflow error
    udrflwd: in std_logic; -- underflow error
    quotient: in signed(OP_SIZ-1 downto 0);
    -- outputs
    iteration: inout integer range 0 to 9;
    i: out std_Logic; --imaginary
    rdy: out std_logic; -- data ready
    err: out std_logic; -- error, input out of range
    ovrflw: out std_logic; -- multipliers or dividers
    udrflw: out std_logic; -- division underflow
    rt: out signed((1+INT_SIZ+FRAC_SIZ)-1 downto 0)
);
end nthRoot;

architecture RTL of nthRoot is

-- signals
signal \Xi\,\Xi+1\,diff,last_diff: signed(OP_SIZ-1 downto 0) := (others=>'0');
signal sr,op,t1,t2,t3,s1: signed(OP_SIZ-1 downto 0) := (others=>'0');
signal busy: std_logic := '0';
signal nth: natural range 2 to 5 := 2;
signal sign: std_logic := '0';
signal src: natural range 0 to OP_SIZ/3 := 0;

alias grs is op(3 downto 0); -- guard, round, and two sticky

-- state machine enumeration list
type sm_def is
(
    RESET,
    START_EXTR,
    PRESCALE,
    PRESCALE2,
    MAKE_GUESS,
    MAKE_GUESS2,
    MAKE_GUESS3,
```

```
    ALGORITHM,
    ALGORITHM2,
    ALGORITHM3,
    ALGORITHM4,
    ALGORITHM5,
    ROUND,
    POSTSCALE,
    COMPLETE,
    ERROR,
    OVERFLOW,
    UNDERFLOW,
    CONVERT_SIGN
);
signal state, ret_state: sm_def := RESET;

begin
--
--   nth root algorithm state-machine
--
process(rst,clk)
begin
    if(rst='1') then

            -- working registers
            \Xi\ <= (others=>'0');
            \Xi+1\ <= (others=>'0');
            diff <= (others=>'0');
            last_diff <= (others=>'0');
            sr <= (others=>'0');
            op <= (others=>'0');
            t1 <= (others=>'0');
            t2 <= (others=>'0');
            t3 <= (others=>'0');
            s1 <= (others=>'0');

            -- signals to external modules
            startm <= '0';
            multiplier <= (others=>'0');
            multiplicand <= (others=>'0');
            startd <= '0';
            divisor <= (others=>'0');
            dividend <= (others=>'0');
            startp <= '0';
            power <= 1;
```

```vhdl
        base <= (others=>'0');

        -- flags
        busy <='0';
        err <='0';
        ovrflw <='0';
        udrflw <='0';

        -- misc
        i <= '0';
        nth <= 2;
        sign <= '0';
        iteration <= 0;
        rt <= (others=>'0');
        src <= 0;


        -- states
        state <= RESET;
    elsif rising_edge(clk) then
        -- 1 clock signals
        startm <= '0';
        startd <= '0';
        startp <= '0';


        --
        --   state machine body
        --
        case state is
            when RESET =>
                state <= START_EXTR;
            -- wait for signal to start extraction
            when START_EXTR =>
                --account for extra guard, round, and sticky bits
                op <= resize((rcd&"0000"),OP_SIZ);
                if(extr = '1') then
                    -- convert negative operands
                    if(rcd(rcd'high) = '1') then
                        sign <= '1';
                        state <= CONVERT_SIGN;
                        ret_state <= PRESCALE;
                    else
                        sign <= '0';
                        state <= PRESCALE;
```

```
        end if;
        -- root order value
        nth <= n;
        i <= '0';
        busy <= '1';
        err <= '0';
        ovrflw <= '0';
        udrflw <= '0';
        iteration <= 0;
        src <= 0;
        last_diff <= (others=>'0');
    end if;
--
-- prescale operand before guess
--
when PRESCALE =>
    sr <= op;   -- put operand in shift register
    -- no integer in format
    if(OP_INT_SIZ = 0 or nth = 2) then
        state <= MAKE_GUESS;
    -- no integer in content
    elsif(op(op'high downto OP_FRAC_SIZ) = 0) then
        state <= MAKE_GUESS;
    -- requires prescale
    elsif(scale_en = '1') then
        state <= PRESCALE2;
    else
        state <= MAKE_GUESS;
    end if;
-- right shift until content visible
when PRESCALE2 =>
    case nth is
        -- index 3 shifts group of 3-bits
        when 3 => if(sr(sr'high downto OP_FRAC_SIZ+16) /= 0) then
                        sr <= "000"&sr(sr'high downto 3);
                        src <= src + 1;
                  else
                        op <= sr;
                        state <= MAKE_GUESS;
                  end if;
        -- index 4 shifts group of 4-bits
        when 4 => if(sr(sr'high downto OP_FRAC_SIZ+10) /= 0) then
                        sr <= "0000"&sr(sr'high downto 4);
                        src <= src + 1;
```

```
                              else
                                  op <= sr;
                                  state <= MAKE_GUESS;
                              end if;
                -- index 5 shifts groups of 5 bits
                when 5 => if(sr(sr'high downto OP_FRAC_SIZ+8) /= 0) then
                                  sr <= "00000"&sr(sr'high downto 5);
                                  src <= src + 1;
                              else
                                  op <= sr;
                                  state <= MAKE_GUESS;
                              end if;
                when others => null;
            end case;
--
-- make initial guess
--
when MAKE_GUESS =>
    -- integer format only -> op/n
    if(FRAC_SIZ = 0) then
        startd <= '1';
        dividend <= op;
        divisor <= root_order(nth);
        state <= MAKE_GUESS2;
    -- fractional content only -> op * n (limited to < 1.0)
    elsif(INT_SIZ = 0) then
        startm <= '1';
        multiplicand <= op;
        multiplier <= root_order(nth);
        state <= MAKE_GUESS3;
    -- integer and fractional portions
    elsif(FRAC_SIZ /= 0 and INT_SIZ /= 0) then
        -- integer content -> op/n
        if(op(OP_SIZ-1 downto OP_FRAC_SIZ) /= 0) then
            startd <= '1';
            dividend <= op;
            divisor <= root_order(nth);
            state <= MAKE_GUESS2;
        -- fractional only content -> op * n (limited to < 1.0)
        elsif(op(OP_FRAC_SIZ-1 downto 0) /= 0) then
            startm <= '1';
            multiplicand <= op;
            multiplier <= root_order(nth);
            state <= MAKE_GUESS3;
```

```
            -- error condition
            else
                state <= ERROR;
            end if;
        -- error condition
        else
            state <= ERROR;
        end if;
        -- check for error but keep result if good
        when MAKE_GUESS2 =>
            if(cmpltd = '1') then
                if(ovrflwd = '1') then
                    state <= OVERFLOW;
                elsif(udrflwd = '1') then
                    state <= UNDERFLOW;
                else
                    \Xi\ <= quotient;
                    state <= ALGORITHM;
                end if;
            end if;
        -- check for error but limit to < 1.0 result if good
        when MAKE_GUESS3 =>
            if(cmpltm = '1') then
                if(ovrflwm = '1') then
                    state <= OVERFLOW;
                else
                    -- limit to max fractional content
                    if(product(OP_SIZ-1 downto OP_FRAC_SIZ) /= 0) then
                        \Xi\ <= (others=>'0');
                        \Xi\(OP_FRAC_SIZ-1 downto 0) <= (others=>'1');
                    else
                        \Xi\ <= product;
                    end if;
                    state <= ALGORITHM;
                end if;
            end if;
    --
    --  primary algorithm
    --
    -- start computing terms 1 and 2
    when ALGORITHM =>
        -- term t1 = ((n-1) * Xi)
        startm <= '1';
        multiplicand <= \Xi\;
```

```
            multiplier <= root_order_minus_1(nth);
            -- term t2 = Xi to the power of n-1
            startp <= '1';
            base <= \Xi\;
            power <= nth-1;
            state <= ALGORITHM2;
-- accept t1 and t2 from multiplier and power function, and
-- start computing term 3
when ALGORITHM2 =>
    if(cmpltm = '1' and cmpltp = '1') then
        -- check for overflow
        if(ovrflwm = '1' or ovrflwp = '1')then
            state <= OVERFLOW;
        elsif(udrflwp = '1') then
            state <= UNDERFLOW;
        else
            -- load term t1
            t1 <= product;
            -- start computing t3 = R/t2
            startd <= '1';
            dividend <= op; -- R
            divisor <= result;-- result is t2
            t2 <= result; -- debug
            state <= ALGORITHM3;
        end if;
    end if;
-- accept t3 and compute sum s1 = t1 + t2, the start computing Xi+1
when ALGORITHM3 =>
    if(cmpltd = '1') then
        -- check for overflow
        if(ovrflwd = '1') then
            state <= OVERFLOW;
        -- check for underflow
        elsif(udrflwd = '1') then
            state <= UNDERFLOW;
        -- start computing Xi+1
        else
            startm <= '1';
            multiplicand <= t1 + quotient; -- quotient is t3, sum is s1
            t3 <= quotient; -- debug
            s1 <= t1 + quotient; -- debug
            multiplier <= root_order_recip(nth);
            state <= ALGORITHM4;
        end if;
```

```vhdl
        end if;
-- accept Xi+1
when ALGORITHM4 =>
    if(cmpltm = '1') then
        -- check for overflow
        if(ovrflwm = '1') then
            state <= OVERFLOW;
        -- check for convergence
        else
            \Xi+1\ <= product;
            state <= ALGORITHM5;
        end if;
    end if;
    -- compute difference between Xi and Xi+1
    diff <= \Xi\ - product;
    -- see if the previous difference is the same but opposite signs
    last_diff <= (\Xi\ - product) + diff;
-- check convergence
when ALGORITHM5 =>
    -- less than or equal to convergence error
    if(diff >= -CONVRG_ERR and diff <= CONVRG_ERR) then
        op <= \Xi+1\;
        state <= ROUND;
    -- if equal but opposite consider converged
    elsif(last_diff = 0) then
        op <= \Xi+1\;
        state <= ROUND;
    else
        \Xi\ <= \Xi+1\;
        state <= ALGORITHM; -- next iteration
    end if;
    -- for measuring performance
    if(iteration = 9) then
        iteration <= 0;
    else
        iteration <= iteration + 1;
    end if;
--
--  round result
--
when ROUND =>
    -- treat even and odd indexes differently
    if(rnd_en = '1') then   -- round only if enabled
        if(nth = 2 or nth = 4) then
```

```
            -- round up even
            if(unsigned(grs) > "1000" or
            (grs = "1000" and op(4) = '1')) then
                sr <= (op(op'high downto 4) + 1)&"0000";
            else
                sr <= op;
            end if;
        else
            -- round away from zero
            if(unsigned(grs) /= 0) then
                sr <= (op(op'high downto 4) + 1)&"0000";
            else
                sr <= op;
            end if;
        end if;
    else
        sr <= op;
    end if;
    state <= POSTSCALE;
    -- post scale operand by left shifting
    when POSTSCALE =>
        if(src > 0)then
            sr <= sr(sr'high-1 downto 0)&'0';
            src <= src - 1;
        else
            -- check for overflow
            if(sr(OP_SIZ-1 downto (1+INT_SIZ+FRAC_SIZ+4)-1) /= 0)
            then
                state <= OVERFLOW;
            -- check for proper sign
            else
                op <= sr;
                -- imaginary or negative root
                case nth is
                    -- wheb index is even
                    when 2 | 4 =>
                    -- sign of radicand
                        if(sign = '0') then
                            state <= COMPLETE;
                        else
                            i <= '1'; -- imaginary sign
                            state <= COMPLETE;
                        end if;
                    -- when index is odd
```

```
                    when 3 | 5 =>
                        -- sign of radicand
                        if(sign = '0') then
                            state <= COMPLETE;
                        else
                            state <= CONVERT_SIGN;
                            ret_state <= COMPLETE;
                        end if;
                end case;
            end if;
        end if;
    -- extraction complete
    when COMPLETE =>
        rt <= op((1+INT_SIZ+FRAC_SIZ+4)-1 downto 4);
        busy <= '0';
        state <= START_EXTR;
    --
    --    subroutines
    --
    when ERROR =>
        err <= '1';
        state <= COMPLETE;
    when OVERFLOW =>
        ovrflw <= '1';
        state <= COMPLETE;
    when UNDERFLOW =>
        udrflw <= '1';
        state <= COMPLETE;
    -- only operates on internal operand op
    when CONVERT_SIGN =>
        op <= (not op) + 1;
        state <= ret_state;
    when others =>
        state <= RESET;
    end case;
    end if;
end process;

rdy <= '1' when busy = '0' and (extr = '0') else '0';


end RTL;
```

```
-------------------------------------------------------------
--
--   nthRootPackage.vhd (Newton-Raphson)
--
-------------------------------------------------------------
library IEEE;
use IEEE.std_logic_1164.all;
use IEEE.numeric_std.all;
use IEEE.math_real.all;
use work.ConversionPackageV2.all;

--   package header
package nthRootPackage is
    --
    --   function declaration
    --
    -- functions that return fixed-point constants
    function root_order(n: natural range 2 to 5) return signed;
    function root_order_minus_1(n: natural range 2 to 5) return signed;
    function root_order_recip(n: natural range 2 to 5) return signed;


    --
    --   Fixed-point data formats
    --
    -- input/output fixed-point format (default Qm.n = Q31.32)
    constant INT_SIZ: natural := 31;
    constant FRAC_SIZ: natural := 32;
    -- internal operand scaling
    -- -- minimum of 3 integer portion still required for internal operands
    constant INT_EXT: natural := 3;
    constant OP_INT_SIZ: natural := INT_SIZ + INT_EXT;
    -- include guard, round, and two sticky bits
    constant OP_FRAC_SIZ: natural := FRAC_SIZ+4;
    constant OP_SIZ: natural := 1+OP_INT_SIZ+OP_FRAC_SIZ;
    constant CONVRG_ERR: natural := 1; -- 1-bit error tolerance is default
    -- simple order constant
    constant ORDER_2: signed (OP_SIZ-1 downto 0) :=
        RealToQmn(2.0, OP_INT_SIZ, OP_FRAC_SIZ, TZERO);
    constant ORDER_3: signed (OP_SIZ-1 downto 0) :=
        RealToQmn(3.0, OP_INT_SIZ, OP_FRAC_SIZ, TZERO);
    constant ORDER_4: signed (OP_SIZ-1 downto 0) :=
        RealToQmn(4.0, OP_INT_SIZ, OP_FRAC_SIZ, TZERO);
    constant ORDER_5: signed (OP_SIZ-1 downto 0) :=
```

```
    RealToQmn(5.0, OP_INT_SIZ, OP_FRAC_SIZ, TZERO);
    -- index order constant - 1 for n-1 in expressions
    constant ORDER_2_MINUS_1: signed (OP_SIZ-1 downto 0) :=
RealToQmn(1.0, OP_INT_SIZ, OP_FRAC_SIZ, TZERO);
    constant ORDER_3_MINUS_1: signed (OP_SIZ-1 downto 0) :=
RealToQmn(2.0, OP_INT_SIZ, OP_FRAC_SIZ, TZERO);
    constant ORDER_4_MINUS_1: signed (OP_SIZ-1 downto 0) :=
RealToQmn(3.0, OP_INT_SIZ, OP_FRAC_SIZ, TZERO);
    constant ORDER_5_MINUS_1: signed (OP_SIZ-1 downto 0) :=
RealToQmn(4.0, OP_INT_SIZ, OP_FRAC_SIZ, TZERO);
    -- index order reciprocal  for 1/n in expressions
    constant ORDER_2_RECIP: signed (OP_SIZ-1 downto 0) :=
RealToQmn(1.0/2.0, OP_INT_SIZ, OP_FRAC_SIZ, TZERO);
    constant ORDER_3_RECIP: signed (OP_SIZ-1 downto 0) :=
RealToQmn(1.0/3.0, OP_INT_SIZ, OP_FRAC_SIZ, TZERO);
    constant ORDER_4_RECIP: signed (OP_SIZ-1 downto 0) :=
RealToQmn(1.0/4.0, OP_INT_SIZ, OP_FRAC_SIZ, TZERO);
    constant ORDER_5_RECIP: signed (OP_SIZ-1 downto 0) :=
RealToQmn(1.0/5.0, OP_INT_SIZ, OP_FRAC_SIZ, TZERO);

end;

--   package body
package body nthRootPackage is


    -- n in fixed-point
    function root_order(n: natural range 2 to 5) return signed is
    variable order: signed(OP_SIZ-1 downto 0) := (others=>'0');
    begin
            case n is
                when 2 => order := ORDER_2;
                when 3 => order := ORDER_3;
                when 4 => order := ORDER_4;
                when 5 => order := ORDER_5;
            end case;
            return(order);
    end function;
    -- n-1 in fixed-point
    function root_order_minus_1(n: natural range 2 to 5) return signed is
    variable order: signed(OP_SIZ-1 downto 0) := (others=>'0');
    begin
            case n is
                when 2 => order := ORDER_2_MINUS_1;
```

```
                when 3 => order := ORDER_3_MINUS_1;
                when 4 => order := ORDER_4_MINUS_1;
                when 5 => order := ORDER_5_MINUS_1;
            end case;
            return(order);
    end function;
    -- 1/n in fixed-point
    function root_order_recip(n: natural range 2 to 5) return signed is
    variable order: signed(OP_SIZ-1 downto 0) := (others=>'0');
    begin
            case n is
                when 2 => order := ORDER_2_RECIP;
                when 3 => order := ORDER_3_RECIP;
                when 4 => order := ORDER_4_RECIP;
                when 5 => order := ORDER_5_RECIP;
            end case;
            return(order);
    end function;

end;
```

9 Carry-Save Adders

Because several different variants of carry-save adders are used throughout this book a short chapter is dedicated to the subject. It is referenced from other chapters as well as other books in this series.

The first advantage of using carry-save form is that the carry propagation of an operand or intermediate register is postponed until later when the two's complement representation is needed, typically after the algorithm has completed. Beforehand it is maintained with two components, a sum and carry of equal bit-width. The second advantage is the propagation delay is limited to only a single full-adder or a single LUT (lookup table) within an FPGA logic block.

The simplest carry-save adder is shown in Figure 24 (the most basic version would not have either a carry in or carry out). A 3-to-2 carry-save adder has three inputs and two outputs. Inputs can be two's complement or a combination of carry-save form and two's complement, but the outputs are always in carry-save form.

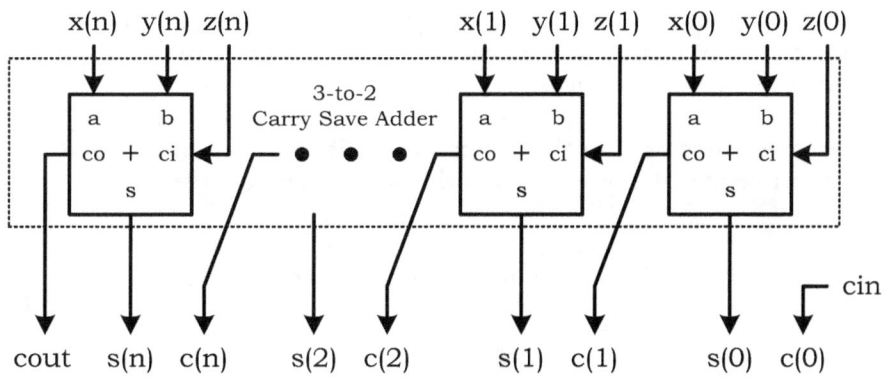

Figure 24

Figure 25 is drawn slightly different and has 2 levels of carry-save adders supporting a forth input. This is a 4-to-2 carry-save adder and has a delay of two full adders in series or two LUTs plus net delays. As with the 3-to-2 adddder inputs can be a combination of carry-save components and two's complement operands.

Note: *there are two carry in inputs and a single carry out. An operand can be subtracted by inverting its input and setting one of the carry in bits true.*

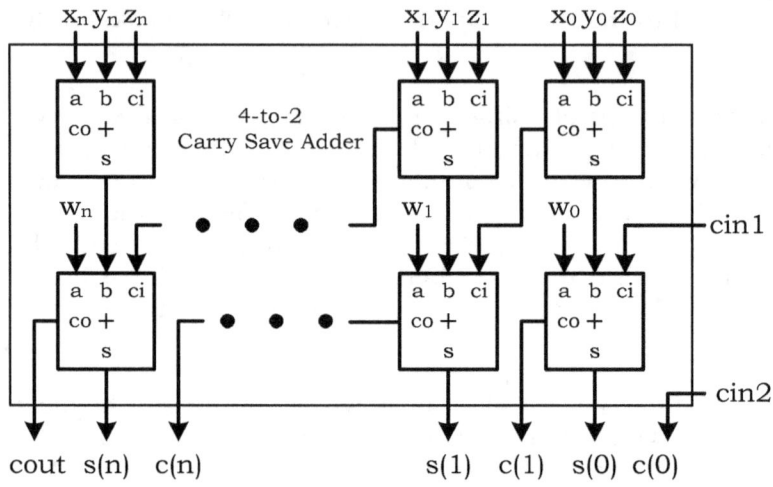

Figure 25

Figure 26 has 3 levels of 3-to-2 carry-save adders and can add 6 operands, combining carry-save form and two's complement and outputting carry-save form.

Three levels means that the propagation delay is equal to three full-adders in series or three LUTs and interconnections delays.

Note: *there are four carry inputs and one carry output.*

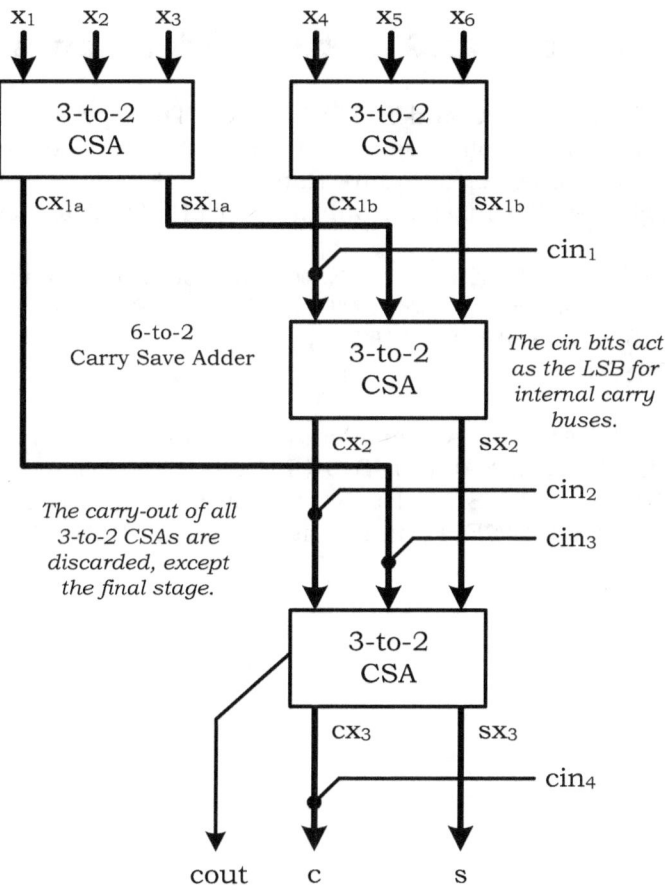

Figure 26

10 Permissions and Acknowledgements

[1] MILOS ERCEGOVAC and TOMAS LANG, "Digital Arithmetic" Morgan Kaufmann Publishers, an imprint of Elsevier Science, Copyright 2004, chapter 6 Square Root by Digit Recurrence" pp. 331 through 365. Permission from Elsevier LN: 4340390823590.

[2] HONG PEND, "Algorithms for Extracting Square Roots and Cube Roots", Computer Arithmetic (ARITH), 1981 IEEE 5th Symposium, pp. 124-126. Copyright 2012 IEEE Permissions LN: 4359020838341.

[3] NAOFUMI TAKAGI, "A Digit-Recurrence Algorithm for Cube Rooting", IEICE TRANSACTIONS ON FUNDAMENTALS OF ELECTRONICS, VOL.E84-A, NO.5, pp. 1309-1314, MAY 2001, Copyright (c)2016 IEICE, Permissions Number: 18RA0033.

11 APPENDIX

The source code that follows supports conversions between real and fixed-point numbers, which exceed the 32-bit limits.

```vhdl
--
--  File: ConversionPackageV2.vhd
--
--  Changes from initial version:
--  1). Some synthesis tools will not allow parameters m or n
--  to be 0. Changes were made to support 0 values.
--  2). The RealToQmn function now allows for rounding.
--  3). Problem fixed on RealToQmn when integer > 30 bits
--
library IEEE;
use ieee.std_logic_1164.all;
use ieee.numeric_std.all;
use work.all;

--  package header
package ConversionPackageV2 is
    -- note: all numbers are processed as positive
    type r_def is
    (
        TZERO, -- truncate
        AZERO, -- round if any lower bits
        NEVEN -- .5 remainder and lsb even
    );

    function RealToQmn(r: real; m,n: integer; rnd: r_def) return signed;
    function QmnToReal(qmn: signed; m,n: integer) return real;

end;

--  package body
package body ConversionPackageV2 is


    --
    --  Convert real number to Qmn fixed-point number
    --
    --  m - number of integer bits excluding the sign (no limit)
    --  n - number of fractional bits (no limit)
    --  rnd - type of rounding operation
```

```vhdl
function RealToQmn(r: real; m,n: integer; rnd: r_def) return signed is
variable rx,tmp: real := 0.0;
variable int: unsigned(m-1 downto 0) := (others=>'0');
variable frac: unsigned(n-1 downto 0) := (others=>'0');
variable qmn: signed((m+n) downto 0) := (others=>'0');
begin
    -- convert to positive
    if(r < 0.0) then
        rx := (r * (-1.0));
    else
        rx := r;
    end if;
    -- compute integer portion
    if(m > 0)    then
        for i in m-1 downto 0 loop
            -- integer size limit
            if(i < 31) then
                -- subtract binary equivalent
                if((rx - real(2**i)) >= 0.0) then
                    rx := rx - real(2**i);
                    int(i) := '1';
                end if;
            -- multiply up limit by 2
            else
                tmp := real(2**30);
                --for j in (m-1) downto 31 loop
                for j in i downto 31 loop
                    tmp := tmp * 2.0;
                end loop;
                -- subtract binary equivalent
                if((rx - tmp) >= 0.0) then
                    rx := rx - tmp;
                    int(i) := '1';
                end if;
            end if;
        end loop;
    end if;
    -- compute fractional portion (remaining in rx)
    if(n > 0) then
        for i in n-1 downto 0 loop
            -- integer size limit
            if((n-i) < 31) then
                -- subtract reciprocal
                tmp := (1.0/real(2**(n-i)));
```

```
                if((rx - tmp) >= 0.0) then
                    rx := rx - tmp;
                    frac(i) := '1';
                end if;
            -- divide down limit by 2
            else
                tmp := 1.0/real(2**30);
                for j in 31 to (n-i) loop
                    tmp := tmp / 2.0;
                end loop;
                -- subtract reciprocal
                if((rx - tmp) >= 0.0) then
                    rx := rx - tmp;
                    frac(i) := '1';
                end if;
            end if;
        end loop;
        -- adjust remainder for rounding
        for i in 1 to n loop
            rx := rx * 2.0;
        end loop;
end if;


-- construct final fixed-point number
if(m = 0 and n = 0) then
    report "m and n parameters cannot both be zero!" severity failure;
elsif(m = 0) then
    qmn := signed('0'&frac);
elsif(n = 0) then
    qmn := signed('0'&int);
else
    qmn := signed('0'&int&frac);
end if;


-- round according to parameter
if(rnd = AZERO and rx > 0.0) then
    qmn := qmn + 1;
elsif(rnd = NEVEN and (rx > 0.5 or (rx = 0.5 and qmn(qmn'low) = '1'))) then
    qmn := qmn + 1;
else
    null; -- truncate
end if;


-- convert to negative if needed
```

```vhdl
    if(r < 0.0) then
        qmn := (not qmn) + 1;
    end if;

    return(qmn);

end function;
--
--   Convert Qmn fixed-point number to real
--
--   m - number of integer bits excluding the sign (no limit)
--   n - number of fractional bits (no limit)
--
function QmnToReal(qmn: signed; m,n: integer) return real is
variable qmnx: signed(qmn'high downto qmn'low) := (others=>'0');
variable r,tmp: real := 0.0;
begin
    -- convert to positive number
    if(qmn(qmn'high) = '1') then
        qmnx := (not qmn) + 1;
    else
        qmnx := qmn;
    end if;
    -- compute integer portion
    if(m > 0) then
        -- add corresponding power of 2
        for i in qmnx'high-m to qmnx'high-1 loop
            if(qmnx(i) = '1') then
                -- integer size limit
                if((i-n) < 31) then
                    r := r + real(2**(i-n));
                -- multiply up limit by 2
                else
                    tmp := real(2**30);
                    for j in (i-n) downto 31 loop
                        tmp := tmp * 2.0;
                    end loop;
                    r := r + tmp;
                end if;
            end if;
        end loop;
    end if;
    -- compute fractional portion
    if(n > 0) then
```

```
           -- add corresponding power of two reciprocal
           for i in qmnx'high-1-m downto qmnx'low loop
               if(qmnx(i) = '1') then
                   -- integer size limit
                   if((n-i) < 31) then
                       r := r + (1.0/real(2**(n-i)));
                   -- divide down limit by 2
                   else
                       tmp := 1.0/real(2**30);
                       for j in 31 to (n-i) loop
                           tmp := tmp / 2.0;
                       end loop;
                       r := r + tmp;
                   end if;
               end if;
           end loop;
       end if;
       -- restore sign
       if(qmn(qmn'high) = '1') then
           r := (r * (-1.0));
       end if;

       return(r);

   end function;

end;
```